TYRES IN MINING AND ALLIED SECTORS—STATUS AND OUTLOOK

TYRES IN MINING AND ALLIED SECTORS—STATUS AND OUTLOOK

Editor

A.K. Basu
Professor, Department of Mechanical Engineering
and Mining Machinery Engineering
Indian School of Mines
Dhanbad, India

ALLIED PUBLISHERS PVT. LTD.

New Delhi • Mumbai • Kolkata • Lucknow • Chennai
Nagpur • Bangaluru • Hyderabad • Ahmedabad

TYRES IN MINING AND ALLIED SECTORS

Editor
Prof. A.K. Basu

Organised by:

Indian School of Mines
Dhanbad – 826004
India

ISBN:

Typeset by Shri Tapas Dasgoswami and Shri V.V.V. Rao at Indian School of Mines, Dhanbad and Published by Sunil Sachdev and printed by Ravi Sachdev at Allied Publishers Pvt. Ltd., (Printing Division), A-104, Mayapuri Phase-II, New Delhi-110064.

Foreword

The mobile earthmoving machines deployed in mining and allied industries play a significant role in achieving the production target. The tyres are the vital components in the operation of such mobile machines. Consequently, the different aspects of tyres like manufacturing, failure, maintenance and reuse of tyres have attracted the attention of the user industries. Keeping these in view, the Indian School of Mines University, Dhanbad, organized the National Seminar on "**Tyres in Mining and Allied Sectors—Status and Outlook.**"

This National Seminar has attracted many authors from the tyre manufacturing companies, mining and allied sectors, academic institutions, R&D institutes etc. In all, twenty two technical papers have been selected for presentation and publication in the Seminar Volume. This volume contains the technical papers dealing with different aspects of tyres including the latest developments and future challenges.

It is hoped that this Seminar Volume would serve as a reference book for the postgraduate students, academicians, practicing engineers and other professionals connected with this subject.

T. Kumar
Director
ISM University
Dhanbad-826004

Preface

The mining and allied industries are now fully geared to meet the challenges of the user industries commensurate with the overall industrial growth in the country. In this context, the earthmoving machines play a dominant role to maximize the production capacity. The tyres are the critical elements contributing to the operation and maintenance cost of mobile earthmoving machines. Selection, application and management of tyres are the important issues in relation to day-to-day operation of the tyre mounted machines deployed in these industries. The problems associated with the use of Off-the-Road (OTR) tyres are largely different from those linked to other transportation systems.

In view of the above, the first National Seminar on **"Tyres in Mining and Allied Sectors—Status and Outlook"** was organized by the Indian School of Mines, Dhanbad. Delegates from tyre manufacturing companies, mining and allied industries, educational institutions, R&D laboratories etc. have attended the seminar. The Seminar Volume contains twenty two numbers of selected technical papers to cover the various aspects of tyres like design and manufacturing, testing, failure and maintenance and the disposal of scrap tyres.

I express my sincere thanks to the Chairman, Director and Registrar of the Indian School of Mines, Dhanbad for the assistance extended to organize this seminar. I am specially thankful to the Seminar Organising Secretaries and Treasurer for their whole-hearted assistance in executing various activities of the seminar as well as for publishing this seminar volume. Further, I am thankful to all the faculty members, officers, staff members and external members, who were associated in organizing this seminar and in preparing the materials of this volume. I also thank the sponsorers and the delegates for the support extended in relation to this seminar.

A.K. Basu

Contents

Foreword ... *v*
T. Kumar

Preface ... *vii*

1. Study of OTR Tyres in Narwapahar Underground Mine 1
R.P. Sengupta and Rajesh Kumar

2. Maintenance and Repair of OTR Tyres in Opencast Mines 11
P.T. Rao

3. Maintenance Efforts for OTR Tyre Life Improvement—A Case Study from Opencast Mines 25
R.K. Darad, R. Ghosh and K. Mukherjee

4. Selection and Maintenance of OTR Tyres—Failure Analysis and Remedial Measures 38
N. Thangavelu

5. Tyres in Mining Industry—Repair and Maintenance 51
J.N. Singh, R.L. Rai, S.N. Singh, A.K. Pandey, Sandeep Shukla and R.N. Tripathi

6. Off-Highway Tyres in Opencast Mines 58
Deepak Vidyarthi

7. Earthmover Tyres in Bailadila Iron Ore Project (Deposit No. 14 & 11C) 65
V.K. Jain

8. Study of Tyres in West Bokaro Mines—An Overview 69
Sureshwar Singh and P.N. Tiwary

9. Dumper Tyre Failures on Openpit Haulage Routes: A Case Study 78
S.S. Gupte and I.L. Muthreja

10. How to Get Better Tyre Life for Economic Earthmoving 83
U. Das and W.P. Mahapatra

11. The Use of Off-Highway Tyres, Its Importance and Earthmover Management System (EMS) 94
S.S. Gupte and I.L. Muthreja

12. Status of Abrasion Resistant Tyres for Mining Industry: A Review ... 100
Samir K. Pal and Kaushik Pal

13. Thermogravimetric Studies of Different Tyres and Tubes ... 115
Emranuzzaman, M.K. Choudhary, Jagdish, B.M.P. Pingua and Nabiullah

14. Oxidative Degradation of Tyres ... 124
Tanushree Chowdhury and N.M. Mishra

15. Disposal of Used OTR Tyres ... 128
M.S. Waldia and S. Banerjee

16. OTR Tyres—Status and Outlook ... 133
Jacob George

17. Application of OTR Tyres ... 141
R. Sengupta

18. Use of OTR Tyres in Bharat Coking Coal Limited ... 146
V.B. Singh

19. Attainment of Agility through Just-in-Time Approach in a Premier Truck-Tyre Manufacturing Unit ... 151
S. Chattopadhyay, A.K. Basu, Indira Prasanna and Reena Bajaj

20. A Novel PC-Based Cure Monitoring Process of OTR Tyres ... 161
P.K. Sadhu, S. Chattopadhyay and D.K. Mittra

21. Built-in-Diagnosis to Future Technology for Micro-Manufacturing—A Case Study from Tyre Making Plant ... 168
Somnath Chattopadhyay, Surender Kumar and Pankaj Mishra

22. Development of OTR Tyre Mould through Reverse Engineering ... 178
Vinay Sharma, S. Chattopadhyay, P.S. Mukherjee, Reena Bajaj and Indira Prasanna

Index 182

Study of OTR Tyres in Narwapahar Underground Mine

R.P. Sengupta and Rajesh Kumar

Narwapahar Project, Sunder Nagar, Uranium Corporation of India Ltd., Jamshedpur-832107

1. Introduction

Narwapahar mine of Uranium Corporation of India Ltd. is the only underground mine in India, where the fully trackless mining system is in operation. Various trackless equipment required for drilling, loading, hauling and service back-up were introduced a decade ago and these have been working successfully to meet the targeted production of mines.

While the pneumatic rubber tyres are widely used for different operations, the tyre life plays an important role in the trackless system of underground mining to meet the requirements of effective equipment operation, availability of equipment and productivity. The working conditions in underground mines are relatively more severe for tyres. This is mainly due to water seepage, restrictions in roadways, limited availability of space for operations, etc. Thus, the selection of correct size and type of tyre for different applications in a cost effective manner is the challenging task for the engineers engaged in mining operations.

The development work of Narwapahar Mines was started in 1991 with the smaller capacity indigenous equipment, while the imported fleet of equipment was engaged 1993 onwards. Initially, the mine was in developing stage and mostly the development work was carried out in levels, drifts, declines etc. until 1995. Since 1995 the emphasis was laid mainly for having production through stops and ramps to achieve the production target.

2. Performance of Tyres

There are 35 nos. of rubber tyred diesel equipment and these are used for mucking, hauling, drilling, service transport etc. Six different sizes of tyres are used in all such equipment as shown in Table 1. It is observed that the tyres required in LHDs are of size 17.5 × 25 and 12.00 × 24 for mucking operation and that in LPDTs are of size 18.00 × 25 and 14.00 × 25 for hauling operation. These tyres are considered to be the most important tyres as they share about 94% of the total cost of tyres used in Narwapahar mines.

The cut and fill stopes were started since 1996 and the contribution of mucking operation was 20% of the total production. The balance 80% production was from the development operation. From the year 1999 the stoping activities increased considerably commensurate with the increased production capacity. Consequently, the scenario was reversed i.e. the share of ore generated from stopes was raised to about 70 to 80%. Although there are a number of tyres of different types and sizes as used in various trackless equipment, it may be noted from Fig. 1 that the LHDs and LPDTs share about 62% and 32% of the total tyre cost respectively at this project.

Table 1: Population of Trackless Equipment at Narwapahar Mine

Sl. No	*Description*	*Capacity*	*No. of equipment used*	*Type used by manufacturer*	*Tyre used presently*
1.	LPDT	26 Ton	5	18.00 × 25, 28 Ply E3	18.00 × 25, 32 Ply E4HRL
2.	LPDT	13 Ton	4	14.00 × 25, 20 Ply E3	14.00 × 25 × HD1A
3.	LHD	3.1 m^3	5	17.50 × 25, 20 Ply L56	17.50 R25 × SMD_2^+
4.	LHD	1.7 m^3	5	12.00 × 24, 20 Ply L55	12.00 R24 × SMD_2^+
5.	Drill Jumbo	Twin boom	6	12.00 × R20 steel radial	Same
6.	Service Equipment	-----	10	10.00 × 20, 16 Ply	10.00 R20 × ZHR

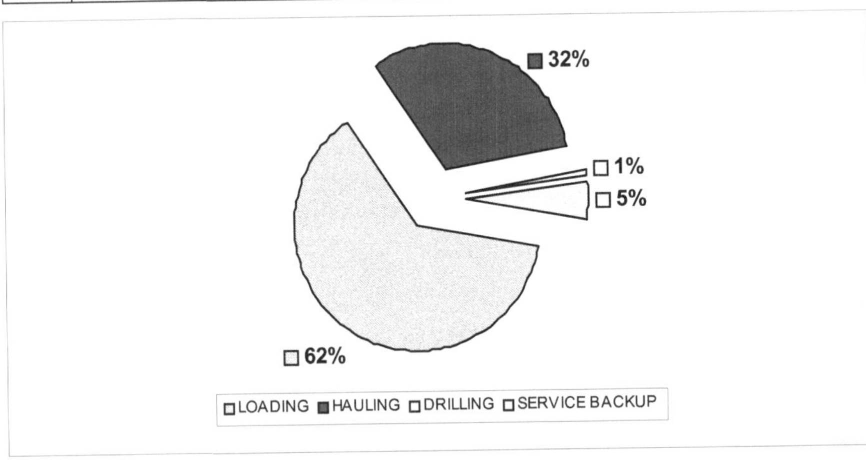

Fig. 1: Tyre Cost in Different Applications

The life of LHD tyre is comparatively lesser as it has to negotiate comparatively more rough roadways and the driving conditions are also more adverse. The abrasive characteristics of the ores, loading and tramming characteristics etc. are also having major influence on tyre performance. The equipment was originally supplied with Cross Ply tyre L5S-sleek type, while the equipment was operating initially for mine development work i.e. drives/drifts. Although the tire performance was not found to be consistent due to the occurrence of frequent side cuts, a trial of cross ply tyres of different make was taken up during this time. However, the tire performance got deteriorated still further when actual production from the stopes was started. As a result, the frequency of failure of tyres and the consequent replacement of tyres had gone up tremendously. The main causes of failure were due to development of side cuts, spinning of tyres during the loading of LHD bucket, negotiating several humps etc. Considerable difficulties were experienced in planning to procure this imported tyre to perform the replacement operation in time. In order to overcome all the above problems, steel radial tyres were used on trial basis for two to three years and finally it was found to be more stable and better both in terms of cost and performance. The major problem of side cuts and the penetration of stones could be obviated with the use of steel wire casing that worked as a shield to protect the tyre. Moreover, the spinning of tyres also got reduced due to the extra contact area of radial type tyre construction. The difference in construction between cross ply and radial tyres has been shown in Fig. 6 & Fig. 7. In radial type construction of tyres, the casing is belted by several plies of steel cords placed at different angles to reinforce and stabilize the trade. The belt has high transverse rigidity, radial flexibility and offer resistance to local deformation. The comparative performance between these two types of tyres i.e. steel radial and Bias types is shown in Fig. 2.

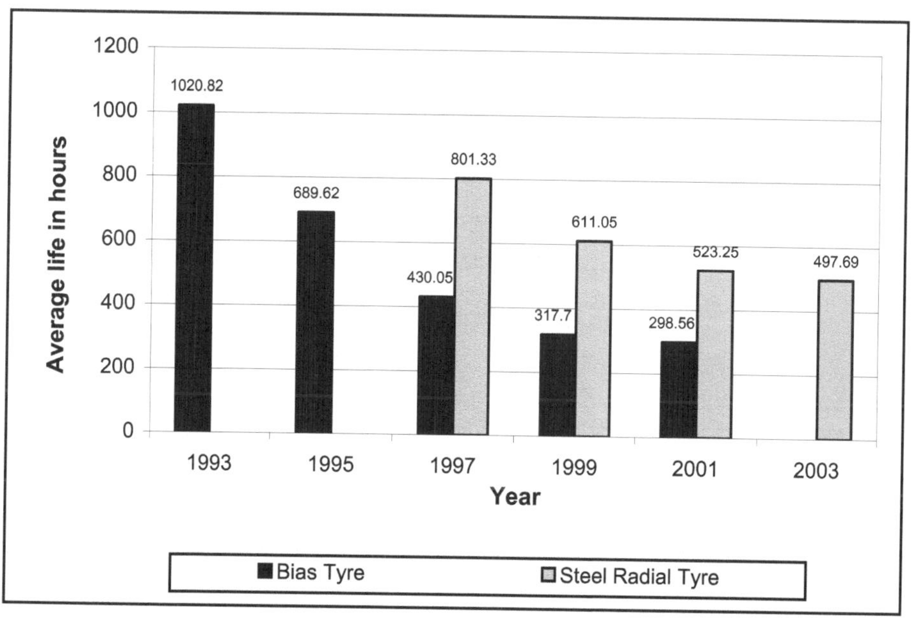

Fig. 2: Comparative Performance of Bias Tyre and Steel Radial Tyre

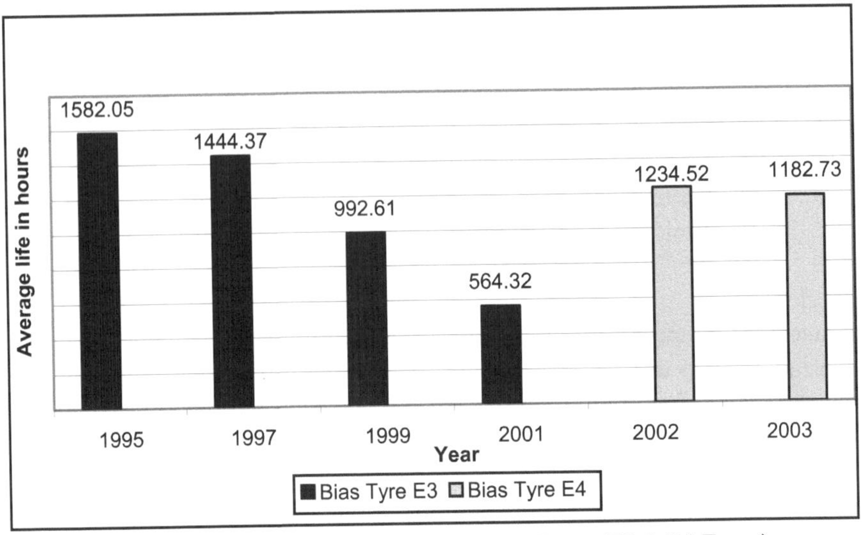

Fig. 3: Comparative Performance of Bias Tyres (E3 & E4 Tyres)

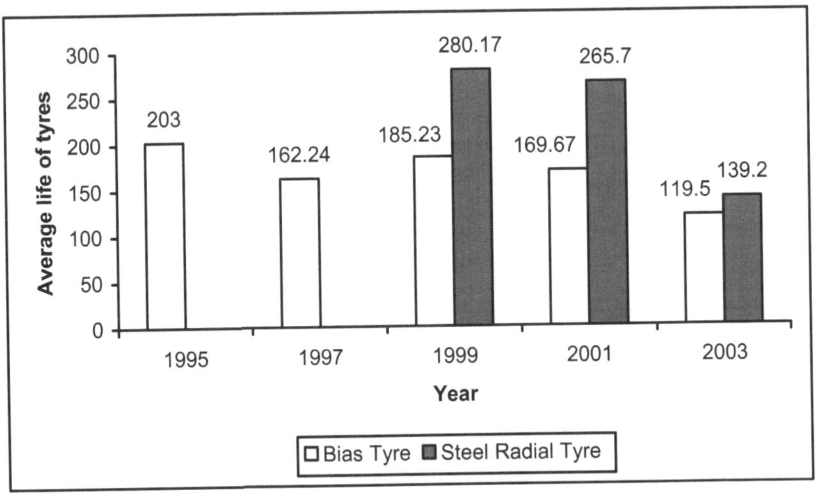

Fig. 4: Comparative performance of Retreading Tyre Size—17.5 x 25

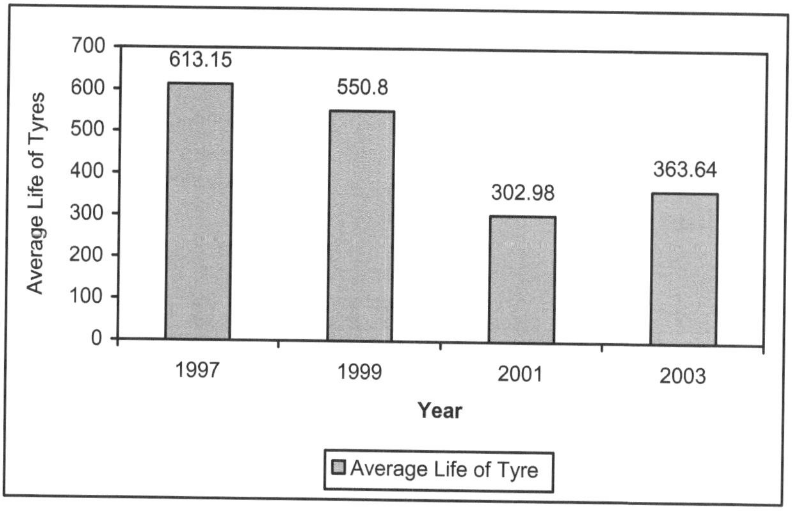

Fig. 5: Performance of Retreading Tyre Size—18.00 × 25

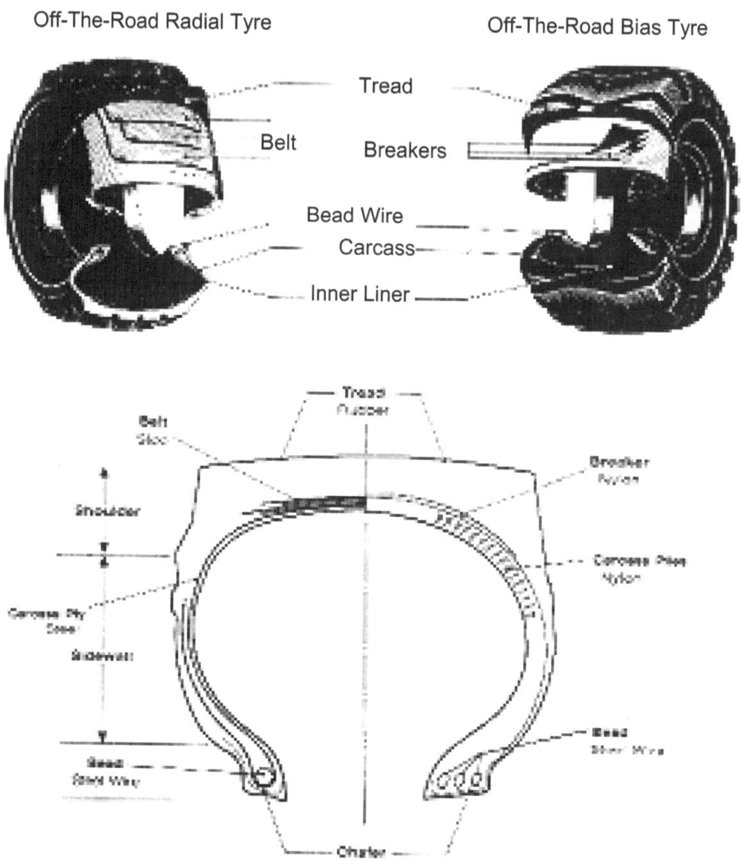

Fig. 6: Type of Tyre (structural diagram)

Hauling	Tread type	Loading	Tread type
E1	Rib	L2	Traction
E2	Traction	L3	Rock
E3	Rock	L4	Rock Deep
E4	Rock deep	L4S	Smooth Deep
E17	Flotation	L5	Rock extra Deep
		L5S	Smooth extra Deep

Different tread depths

Exterior

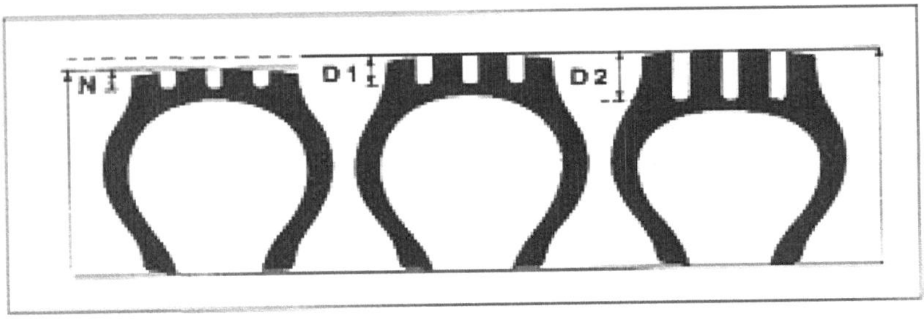

N = Normal (Traction, rock)   D1 = N × 1.5 (Rock, deep tread)   D2 = N × 2.5 (Rock, extra deep tread)

Fig. 7: TRA Classification for Off-the-Road Tyres

The smaller size indigenous LHDs use the radial tyres of 12.00 × 24 size. It has been observed that the performance of radial tyres of this size is better. However, the cost effectiveness of these tyres is yet to be established.

Similarly, the LPDTs (26 T) initially used the tyres of size 18.00 × 25. These LPDTs were deployed in the main drifts and drives, where the loading of LPDT was done directly by LHDs. Most of the time, the LPDTs were hauling the minerals on well maintained roads. However, after opening of the stopes, the loading of LPDTs used to be done through chute, when the tyres of LPDT had to come in contact with accumulated water and spilled over muck during each cycle causing frequent failure due to stone penetration. Over the period of time, the life of tyre got decreased to about 30% of the initial life. Thereafter, high cut resistance, extra deep thread tyre i.e. E4 Type (instead of E3 type) was placed on a trial basis and the performance was found to be much superior as shown in Fig. 3. These are available indigeneously but are marginally costlier.

In case of smaller LPDTs (13 T), the tyres of size 14.00 × 25 and of deep thread type (E3) were used with many constraints mainly due to non-availability of E4 type tyres indigeneously. However, to meet the requirements of waste dumping as well as recycling later for cut and fill stopes, the imported steel radial tyres have been put under trial very recently and the performance of the same is satisfactory, although the cost effectiveness of the same is yet to be ascertained.

3. Factors Affecting Tyre Performance

(a) *Vehicle Speed:* OTR Tyres are desired for heavy duty use and are not meant for high speed of operation. Travelling at excessive speed will increase the internal temperature of tyre that can cause damage to the tyre like bead damage, ply separation and bursting.

(b) *Load:* Tyre size, ply rating, vehicle speed and inflation pressure are the four factors to determine proper load. Overloading of tyres may cause serious tyre damage and reduce its service life as shown in Fig. 9.

(c) *Road Condition:* Negotiation of steep grades may result in rapid tyre wear.

(i) 0–6% grades	:	Excellent, very good life as expected.
(ii) 6–8% grades	:	Good, 10% reduction in life.
(iii) 8–10% grades	:	Bad, 20% reduction in life.
(iv) > 10% grades	:	Very bad, 40% reduction in life.

(d) *Inflation:* A proper inflation would ensure the right tyre deflection and assure the maximum positive traction as shown in Fig. 8.

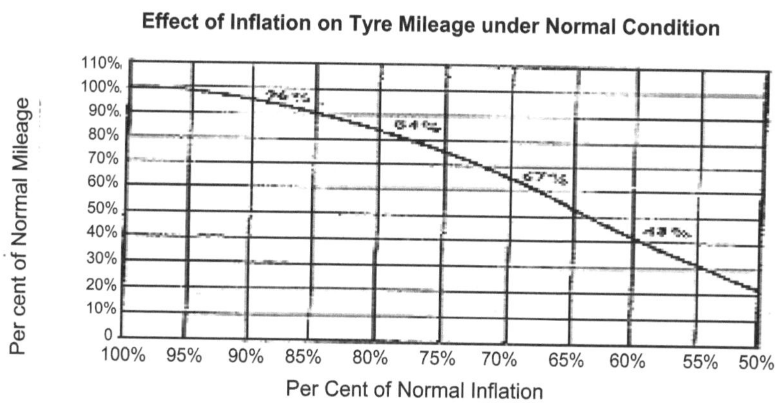

Fig. 8: Effect of Inflation on Tyre Mileage under Normal Condition

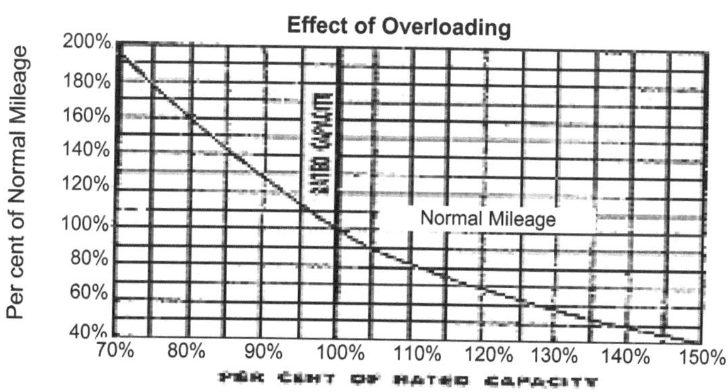

Fig. 9: Effect of Overloading

There may be over-inflation or under-inflation in the use of tyres. The following effects are generally observed as below:

(i) Over-inflation
   (a) Rapid wear at the center of tyre.
   (b) Reduced contact leads to reduced traction effect.
   (c) Excessive pressure may damage the tyre bead.
   (d) Increased tension in cord and tread result in the development of cut and damage due to impact.

(ii) Under-inflation
   (a) Excessive tyre deflection.
   (b) Increased side wall movement and thus developing higher internal temperature.
   (c) Decreases traction between rim and tyre. This may result in leakage development in tubeless tyres.

4. Field Observations

Water logging and muck spillage on roadways are the main causes for poor performance of tyres. Sufficient care has been taken by engaging the motor grader regularly for the maintenance of roadways. Periodic cleaning of side drains is ensured to avoid any water accumulation in the main haulage road. When water is in contact with rubber, it makes the tyre treads softer and may increase the possibility of stone piercing rate two fold than that in dry condition. However, there is limitation to maintain clean and dry condition of roadways in cut and fill stopes mainly due to its operational constraints. Proper care is taken to minimize such problems.

After the change over to steel radial tyres on LHDs and the E4 type tyres on LPDTs, the performance could be stabilized.

5. Retreading of Tyres

There is definite advantage of saving in direct cost with the use of retreated tyres. In order to get the tyre retreated, it has to be dismantled from the equipment before the damage of ply and casing. About 10–15% of full life is sacrificed for retreading. A complete techno-economic analysis is needed for special and costly tyres with regard to their retreading.

In underground mining application, it has been observed that the retreaded tyres are cost effective in haulage application, but these are not reliable as well as not cost effective in loading application. This is due to poor average life and frequent dismantling of tyres would incur indirect losses. Retreading of LHD tyres are not taken into account for regular use as the failure of tyres is experienced due to following reasons:

- Bonding of retreaded rubber with the casing would become loose later on.
- Cases of cut/penetration are very high.

The LHD tyres of size 17.50 × 25 are imported and these are of nonstandard size. Hence, arrangement has been made to have some retreaded tyres to meet emergency requirement. The performance of retreaded tyres has been shown in Fig. 4 & Fig. 5. Also the tyre performance of size 17.5 × 25 before and after retreading has been shown in Fig. 10.

Fig. 10: Performance of Tyre, Size 17.5 × 25

6. Maintenance of Records

The maintenance department is regularly recording various data for trackless equipment repair using a dedicated computer. Related important information e.g. life of tyres for different size and make, reasons of premature failure etc. are also recorded regularly right from the day of start of equipment operation and this helps to select the best tyres suitable for this mining project.

7. Conclusions

Trackless underground mining started at this project about 30 years ago. However, the tyre manufacturers have now taken up the issue seriously and are now active in improving the quality of tyres to be suitable for underground use. The data presented in this paper are of real life. A regular monitoring of tyres is required as the mining conditions and circumstances may differ from one site to another. However, some key factors as below are to be considered for achieving better performance of tyres and these are very well within the control of the user department.

- Well maintained roadways in underground mines.
- Minimisation of water logging at the working and haulage areas.
- Use of the correct type of steel radial tyres for loading application.

- Use of E4 type of tyres, instead of E3 type, for haulage application of the LPDTs.
- To have regular communication with the tyre manufacturers for improvement of tyre life to be suitable for the underground mines.
- To have regular maintenance of records and analysis of records in relation to the performance of each size and type of different make.

8. Acknowledgement

The authors are thankful to the Chairman & Managing Director, UCIL to present this paper in the seminar based on practical in-house experience.

Maintenance and Repair of OTR Tyres in Opencast Mines

P.T. Rao

Central Workshop, NTS, Barkakhana, Central Coalfields Ltd.
Ramgarh-829103

1. Introduction

OTR tyre is the pneumatic tyre having an annular toroidal shaped inflatable envelope made of rubber, which is reinforced with cord. It is fitted to a metallic rim secured to the undercarriage unit of the Heavy Earth Moving Equipment (HEMM). The pneumatic tyres were introduced during early nineteenth century. However, those tyres were very similar to the bicycle tyres. The tubeless tyres were introduced for the first time in 1947. The principal function of the tyre is to support the weight of the vehicle and this acts as an additional spring in the suspension system to help in converting the engine power to the movement of vehicle. The pneumatic tyre permits cornering on the road at a relatively higher speeds as compared to that with the use of a solid tyre. In HEMM application, the tyre is the capital intensive spare part and the cost of vehicle tyres is about $1/25^{th}$ of the total cost of the machine. The conditions prevailing at the mines, length of the haul road, etc. dictate the design and construction of the tyres. Favourable operating conditions and the maintenance of tyres would drastically improve the productivity of the mines.

2. Classification and Standard

The tyres may be classified in the following ways:

2.1 *Tyre Size*

The tyre size as marked on its sidewall indicates the approximate dimensions of the tyre. Many systems of tyre size representation are in vogue. However, the two most commonly used systems are:

(a) Tyre size 9.00–20″ indicates that the tyre has a section width of approx. 9″ and it is fitted on a rim of 20 inch nominal bead diameter.

(b) In 24R 49E4 2-A tyre designation, 24 indicates the section width in inches, R indicates Radial type of tyre, 49 represents the Rim diameter in inches, E4 shows the TRA code and 2-A is the specification index e.g. cut protected.

2.2 *Construction*

The tyre may be classified by the type and construction of the casing used in assembly. The two types of casing construction are:

(a) *Cross Ply:* The casing angle and the breaker angles are generally equal and are in the range of 40° with the alternate plies running in opposite direction.

(b) *Radial Ply:* All the cords in the casing run in a radial direction and the breaker cords are at a very low angle i.e. almost circumferential.

2.3 *Tread Pattern Design*

The tyres used on the hard and dry roads donot need a patterned tread. However, on wet roads, the water acts as a lubricant between the tread rubber and the road. This helps to reduce the friction considerably and assumes better grip of the tyre on the road.

The pattern on the tread of tyre helps to remove much of the water remaining between the tyre and the road thus improving the grip of tyre. Different patterns are available for use. Basically the pattern aims at providing specific properties for certain services and also for visual impact. Some of the types of patterns are;

a) Circumferential pattern, b) Lugged pattern, c) Square Stud, d) Ribs with side studs, e) Symmetrical transverse, f) Diamond type, g) Improved ribs and side stud, h) Asymmetrical transverse, i) Bar type, and j) Rock type.

The tread design of OTR tyre is mainly of three types viz; Diamond, Bar and Rock.

- *Diamond Type:* This design is used on front rolling wheels. This kind of tread design reduces the rolling resistance, while the grooves help to have a better hold when the tyre moves on slopes.
- *Bar Type:* The bar type is mainly used for the driving wheels and this consists of protruding bars that penetrate into rough ground and helps to improve the effect of traction.
- *Rock Type:* The rock design is specially suited for mining in quarries or for rock fills, where necessary protection against the cutting hazards should be provided. The treads in such designs are wider to provide a better weight distribution and ground contact. This design may be used for both driving and rolling wheels.

2.4 *Tubed and Tubeless Tyre*

Air at a rated pressure is required to be enclosed between the tyre and the rim. In general, this is achieved by using airtight elastic tube or by air sealing between the tyre and the rim without the use of any tube i.e. by providing thick layer of rubber inside the tyre that acts as an air envelop. The tubeless tyres are widely used in HEMM application.

2.5 *Ply Rating*

The ply rating of the tyre usually signifies its strength. It does not indicate the actual number of plies in a tyre, but it provides an index of its strength.

The commonly accepted international standards are TRA (Tyre and Rim Association), USA, ETRO (European Tyre and Rim Technical Organization), Europe, JATMA (Japan Automobile Tyre Manufacaturers Association), Japan. Some of the typical parameters e.g. load capacity, inflation pressure, dimensions of tyre and rim, tube etc. are chosen following these standards.

3. Construction and Components of Tyres

The three main parts of the tyre are—Casing, Tread and Bead

3.1 *Casing*

The casing is made of layers of textile cord fabric surrounded by rubber compound to provide the strength of tyre. The components of a typical casing are:

(a) Plies, (b) Inner Liner, (c) Insulation, (d) Breakers and (e) Chafer.

3.2 *Tread*

It is a relatively thick layer of rubber, which is fitted over the Casing/Breaker assembly. The top portion of the tread that meets the road should be patterned to provide the desired service. The cap compound provides the desired wear and traction characteristics. The sidewall compound is made separately from tread and this requires protective agents to provide for necessary resistance against heat, ageing and flexing. The size, type, ply rating and tyre manufacturers name are generally displayed on the sidewall.

3.3 *Beads*

The bead locks the tyre to the rim. A typical bead is made up of the following components:

- *Bead Coil:* There may be one wire or more and is coated with rubber. The wires are generally made from copper-coated wire.
- *Wrapper:* This layer of textile fabric coated with rubber compound is wrapped around the bead coils to keep the layers of wires in position and provide a good adhesion between the bead coil and the surrounding casing.
- *Apex:* An apex is fitted over the wrapper and it is made of rubber compound. It is generally triangular in shape. This helps in improving the bead shape and avoid the formation of air pockets at the turning up region.
- *Filler:* This is made of textile fabric coated with rubber compound and is fitted over the wrapped bead and apex assembly.

4. Types of Tyre Failure

In opencast mining operation, the kinds of tyre failures generally observed are as below:

1. Normal top tread wear—suitable for retreading;
2. Cut failure;
3. Burst failure;
4. Abnormal top tread wear and exposure of ply;
5. Wear and damage of side wall;
6. Other defects like air leakage, bumps etc.

Different types of abnormal wear and their causes are shown in Table 1.

Table 1: Different Type of Abnormal Wear and their Causes

Wear Modes	*Causes of Wear*
Rapid Wear of Shoulders (Outer edge)	Under inflation, lack of tyre rotation, High speed cornering.
Rapid Wear of Center	Over inflation, lack of tyre rotation.
Cracked Treads	Alternating under/over inflation, high speed driving, over loading.
Wear on Sides	Excessive Camber, lack of tyre rotation, defects in steering and suspension systems.
Feathered Edge	Incorrect toe
Bold Spots	Wheel unbalance.

A case study on failure analysis and performance evaluation of 129 nos. of new cross ply type tyres used at the different mines of Central Coalfields Ltd. is given in Table 2.

Table 2: Defectwise Tyre performance

Defects	*% of Total Failure*	*Max. Life (Hours)*	*Min. Life (Hours)*	*Average Life (Hours)*
1. Top tread wear (suitable for retreading)	34%	7963	2423	5513
2. Through cut	18%	6357	845	3814
3. Burst	36%	14400	2309	7098
4. Sidewall wear/damage	7%	8066	1400	5163
5. Abnormal wear (ply damage)	2%	7169	1992	4097
6. Other type of defects	3%	12697	4689	7685

The above analysis is based on visual inspection. The hidden defects have not been considered in this study i.e. such defects that can be only noticed after the buffing operation is carried out during the retreading process.

However, the actual % failure and the average life of retreadable tyre would be lesser than the figures that are projected in the Table 2.

Table 3: Defectwise Failure Trend at Different Working Hours

Defect	*Working Hours*				
	0-2000	*2001-4000*	*4001-6000*	*6001-8000*	*8001 and Above*
1. Top tread wear (suitable for retreading)	-	25%	29%	42%	4%
2. Through cut	13%	43%	35%	9%	-
3. Burst	3%	10%	25%	25%	38%
4. Sidewall wear/damage	22%	-	34%	33%	11%
5. Abnormal wear (ply damage)	33%	33%	-	33%	-
6. Other types of defects	-	-	25%	50%	25%

The Table 3 clearly shows that there is an increasing trend of normal top tread wear and burst failure with the increase in working hours.

5. Factors Affecting the Tyre Life

5.1 *Air Pressure*

In a pneumatic tyre, it is the contained air that carries the load and not the tyre casing. One of the most important elements of tyre maintenance is to pay due attention to proper inflation of tyre. A correct air pressure under a given load would result in providing the right tyre deflection so as to provide the optimum ground contact of the tread area and produce the proper traction effect. The result of any improper tyre inflation would be as below:

5.1.1 *Under Inflation*

Underinflation of the tyre subjects the sidewall of tyre to excessive flexing. This results in weakening or breaking down of the cord structure. It also results in rapid and uneven wear of the tread. Underinflated tyres loose their load carrying capacity e.g. in a 24.00 R49 tyre, a reduction in air pressure by 5 psi would be equal to a loss of 1130 lbs of load capacity per tyre.

5.1.2 *Over Inflation*

Overinflation of the tyre results in distorting and unduly straining the cord plies—thereby weakening the entire structure of tyre and shortening its life. Cuts and bruises tend to be serious in an overinflated tyre due to greater tension of the tread and cord. It also results in wearing out of the tread at the center and in reducing the traction and the skid resistance.

5.2 *Truck Loading*

Life of a tyre is greatly dependent on the load it has to carry. A 20% overload may result in reducing the life span of tyre by 30% and a 20% underloading may enhance the service life by 50%. Any decrease in speed of the vehicle would reduce the harmful effects of overloading to some extent. Further, any eccentric loading on the truck body would result in uneven wear of the tyres.

5.3 *Speed*

A higher speed of the vehicle would result in quicker wear of the tread of tyre. An increase in average speed from 35 to 45 m.p.h. may reduce the tread life by over 20%. A further increase in speed upto 50 m.p.h. may reduce the tread life by over 30%. It is generally recommended that the maximum speed of Dumper be maintained in the range of 30-40 km/hr, whereas that for Grader as 40 km/hr and that for Dozer/Loader as 8 km/hr.

5.4 *Haul Road*

A good haul road is primarily defined by the road surface performance, good riding criteria etc. which is measured in terms of total surface corrugation as presented in Table 4.

Table 4: Total Surface Corrugation

Sl. No.	*Category of Haul Road*	*Total Surface Corrugation along a Length of 6 mtrs. of Roadway*	*Speed Limit*
i.	Very Good—A	2.0 cm.	40 km/hr or more
ii.	Good—B	Up to 7.0 cm.	40 km/hr
iii.	Tolerable—C	8 – 12 cm.	30 km/hr
iv.	Bad—D	13 – 16 cm.	20 km/hr or lesser

5.4.1 *Effect of Haul Road Condition on Tyre Life*

(a) *Steep Gradient:* Elimination of gradient on haul road is a proven method to enhance the tyre life. However, this is often not possible in real life. The best alternative recommended by OEM is to restrict the gradient within 6% and the gradient should not exceed the predetermined value for a length of 300 to 500 m at one stretch. The rule of thumb is that there should be about one-to-one ratio between weight transfer and grade. Hence, on a 10% grade, the rear wheels are carrying 10% more load than that they would have carried on a level road. This results in severe tyre overloading leading to reduction in the service life of tyre.

(b) *Improper Drainage:* A good drainage is essential for several reasons. Accumulated water on the road surface reduces tyre traction and it exposes the rock embedded in haul roads developing tyre cuts. The wet rubber may develop cuts even double of that from the dry rubber.

(c) *Sharp Curves:* The haul road should be designed with the least possible curves. The curves should be long and sweeping. This not only prevents the loaded truck spillage but reduces lateral tyre stress and tread scrub.

(d) *Inadequate Lighting:* It has been observed that maximum tyre damage takes place at night i.e. between 9 PM and 5 AM. This is caused by poor visibility due to inadequate lighting and sleeping tendency of the drivers.

(e) *Face Condition:* Areas around shovels can produce a large percentage of cut tyres because of: (i) poor clean up of truck loading pad (ii) Bucket spillage, (iii) Overloading of dumper that results in spillage.

(f) *Wheel Spin:* Over watering or rain creates a slippery ground and results in wheel spin. This effect cuts the treads. It is reported that for every revolution of spinning wheel, there may be a loss of 8 to 11 Hrs. of tread service life.

6. Tyre Maintenance

A good tyre maintenance would help to run the dumpers at maximum operating efficiency. The routine tyre maintenance reduces major tyre failures and improves the service life of tyres. The tyres can be maintained effectively by following some simple steps as given below:

6.1 Visual inspection of tyres has to be conducted. Embedded rocks and loose tread rubber are to be moved to prevent any cut separation. An immediate action should be taken against the bubbles or bumps caused by air inflation.

6.2 The air pressure is very critical for tyre life. Hence, a daily hot-inflation checkup for all tyres on the vehicle in operation would be essentially required.

6.3 The dual tyres should be matched using a limit of ¾″ on overall diameter. Bias ply tyre should not be paired with radial tyre. Mixed tyres do not work together to provide the same traction and hauling performance.

6.4 A sound rim-cleaning programme should be established. The air seal on tubeless tyres depends on the clean lines of rims, ‘O’ ring gutters and bead seat bands. Use of valve cap and new ‘O’ rings during each tyre change would provide a positive air seal. Normal diffusion can cause a pressure loss of 2% or lesser in 24 hours. Any tyre with more than 5% pressure drop should be checked to determine the cause of the leakage.

6.5 The dumpers should be taken out of operation when deep cuts or cracks of tyres, distorted rims, missing wheel bolts etc. are noticed.

6.6 Hydraulic or motor oil saturation on tyres should be prevented. This kind of oil related problem would lead to the damage of tyre rubber.

6.7 A weekly programme of cold air pressure check should be established to prevent the tyres from operating in excess of 5 psi under recommendations. Use of dry air would be an asset towards eliminating the rim dust.

6.8 The lifting of tyres by beads using a crane hook or forks should be avoided as this can damage a critical bead area. As an alternative, the tyre should be lifted under the tread by using flat straps.

6.9 The tyres should be stored and shipped in vertical position to avoid stacking; otherwise this can weaken the tyres at the bottom of the stack. Tyre should be placed in a cool and dry spot away from the sunlight to avoid the premature aging.

6.10 Irregular wear of tyres may be caused due to misalignment of the front wheels or worn bearings. The wear on tyres has to be distributed evenly. Certain plans for rotation of tyres amongst the different wheels are recommended after predetermined working hours.

7. Tyre Retreading

7.1 *Definition*

Retreading is the process of replacing the wornout tread of a tyre repeatedly until the casing retains its life. Casing is the important part of a tyre. 70% of the cost of tyre is involved in the casing. Hence, if the tyre is kept in service even after the tread is worn out, it means that 70% of the investment on the tyre is lost. Thus, the only way to get the full value of the investment is to obtain a fresh span of life by retreading the tyre repeatedly.

The casing of a tyre is manufactured at a heavy cost by keeping it above the ply rating with a view to getting not only the first life, but the renewed lives also after retreading it a number of times. If a tyre is used and maintained with care, its tread can be replaced three or four times.

7.2 *Retreading Process*

The retreading process is mainly of two types:

1. Precure Retreading Process (cold retreading)
2. Conventional Retreading Process (Hot retreading)

7.2.1 *Precure Retreading Process*

It is also called cold retreading process. In this process the procured tread rubber is bonded on the tyre at room temperature within a very short time. This process is widely used to retread smaller sizes of tyres like those of transport vehicles and passenger cars.

7.2.2 *Conventional Retreading Process*

It is often called hot retreading process. In this process the curing of tread rubber is done in the mould with steam at a rated pressure. It is used to retread any size of tyres including the OTR tyres. The flow diagram of such retreading process is shown in Fig. 1. The elemental operations involved in the retreading process are as below:

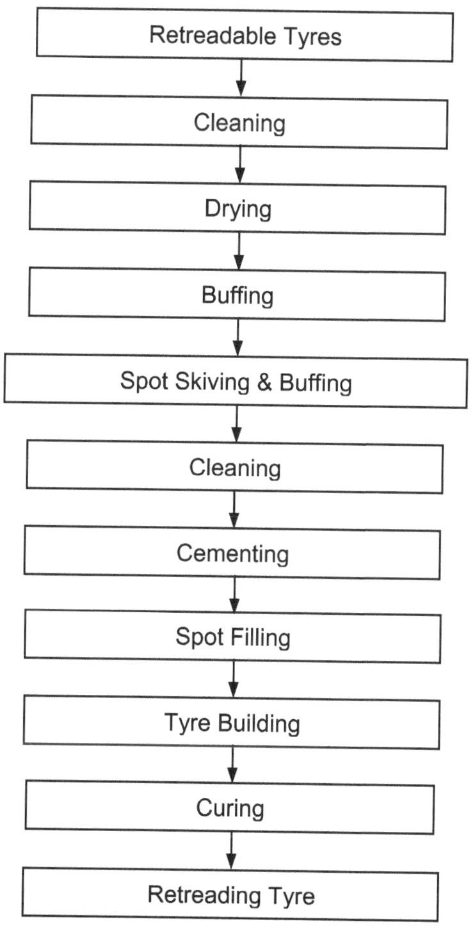

Fig. 1: Flow Diagram for Hot Re-Treading Process

(i) Cleaning : The cleaning of retreadable tyre is first done manually and thereafter by compressed air to remove the dust and other foreign materials.

(ii) Drying : The wet tyre is dried in a drying chamber at the temperature of 50°–55°C for 24 hours to make the tyre free from moisture.

(iii) Buffing : The layer of wornout tread surface is removed by means of Buffing/Detrader Machine using buffing tool. Afterwards, the rough buffed surface is again smoothened by using rasp blade on the same Buffing Machine.

(iv) Spot skiving : Skiving of the cut marks on tyre is done by means of scrapping with the use of knife.

(v) Spot buffing : The cut marks are buffed by hand Buffing Machine.

(vi) Cleaning : The tyre is cleaned by compressed air and by brushing to remove the dust and the rubber chips.

(vii) Cementing : After cleaning of tyre is over, two coats of re-treading cement (used as adhesive) is applied over the buffed surface of the tyre by using a brush. This tyre has to be kept for four hours for getting the cement dried.

(viii) Spot filling : The cut mark spots are filled by cushion gum and tread gum using stitcher.

(ix) Tyre building : Plain tread rubber of required size is then wrapped over the cemented surface with the help of stitching roller fitted on the Detrader Machine. This process is named as tyre building.

(x) Curing : The tyre wrapped with tread rubber is put in the curing mould, in which the tread pattern is already made. Steam is passed from boiler under a pressure of 3.5–5 kg/cm^2 at a temperature of 120°–150°C for six hours in the curing mould. Air at a pressure of 10 kg/cm^2 is also supplied in the tube, which is fitted in the tyre and the same is placed in the mould to press the tyre on the mould so that the tread rubber gets sticked properly on the buffed tyre and a perfect tread pattern can be achieved. After the curing is over, the tyre is taken out of the mould. This tyre is named as the Retreaded tyre.

The equipment required for the hot retreading process are:

(a) Detrader Machine : for buffing and processing of the tread rubber.
(b) Curing Mould : for curing.
(c) Boiler : for steam generation.
(d) Air Compressor : for supplying compressed air required for cleaning.
(e) Chain Pulley Block/Crane : for material handling.
(f) Hand Buffing Machine : for spot buffing.

Materials required:

(i) Tread Rubber of required size of tyre
(ii) Tread gum
(iii) Cushion gum
(iv) Retreading Cement
(v) Petrol
(vi) Tube.

7.3 *Economic Aspect of Tyre Retreading*

A tyre casing when retreaded with proper equipment and materials is often capable of giving the service life as that of a new tyre.

A study on new and retreaded OTR Tyre (size 18.00 × 25 as used in 35 T dumper) life and cost per hour calculated by Tyre Retreading Shop, Central Repair Workshop (CRS), Barkakana on the basis of tyres received for retreading from the mining projects of Central Coalfields Ltd., during the year 2001–02 is presented below:

7.3.1 *New Tyres*

(a) The average working hours of new tyres received at the Tyre Retreading Shop, Central Repair Workshop, Barkakana from the mining areas of Central Coalfields Ltd for the first retreading during the year 2001–02. 3,663
(b) Cost of a new tyre. Rs. 58,000/- (Approx)
(c) Cost per hour of new tyre during the year 2001–02. (a/b) Rs. 15.85

7.3.2 *Retreaded Tyres (Retreaded at the Tyre Retreading Shop, Central Repair Workshop, Barkakana)*

(a) Average working hours of retreaded tyres as received at the Tyre Retreading Shop, Barkakana from the mining projects of Central Coalfields Ltd for 2nd retreading operation during the year 2001–02. 2,635

(b) Cost of retreading a tyre. Rs. 21,500/- (Approx)

(c) Cost per hour of the retreaded tyre during the year year 2001–02. (a/b) Rs. 8.16

It has been observed that:

1. Cost per hour of retreaded tyres is much lesser than that of the new tyres.
2. The average working hour of the retreaded tyres is 72% of that of the new tyres.

8. Cut Repair of Tyres

It is a process by which the cuts and injuries are repaired for a tyre. It is necessary to repair a cut/injury in a tyre, as the 70% value built into a tyre can be harnessed through a number of retreads and hence utilize the tyre to its full money value and design capacity.

The OTR tyres are mostly of tubeless type. The development of a small hole in these tyres would be enough to reject or scrap these tyres. Hence the cut/injury repair in OTR tyres is essential to realize its full life.

A tyre can be repaired by various methods as prevalent in this country and these methods are as below:

(i) Fastening with nuts & bolts.
(ii) Cut piece of used tyres—Gaiter.
(iii) Cord fabric patches.

The 1st and 2nd methods are the old and crude methods of cut repair for tyres and are mostly used for smaller sizes of tyres. The 3rd method is widely used for bigger sizes of tyres including the OTR tyres.

8.1 *Cord Fabric Patches Method*

The flow diagram of cord fabric patches method for through cut repair of OTR tyres is given in Fig. 2. The elemental operations involved in this process are detailed below:

(i) *Cleaning:* The cleaning of a through cut tyre is done manually and by using compressed air jet to remove the dust and other foreign particles/materials etc.

(ii) *Drying:* The wet tyre is dried in a drying chamber at a temperature of about 50°–55° C for 24 hours to make it free from moisture.

(iii) *Through Cut Skiving:* Through cut portion is skived by knife.

(iv) *Through Cut Buffing:* At the outset, the outer surface of cut zone is buffed. Thereafter, the inner surface of the cut portion is buffed by means of hand buffing machine.

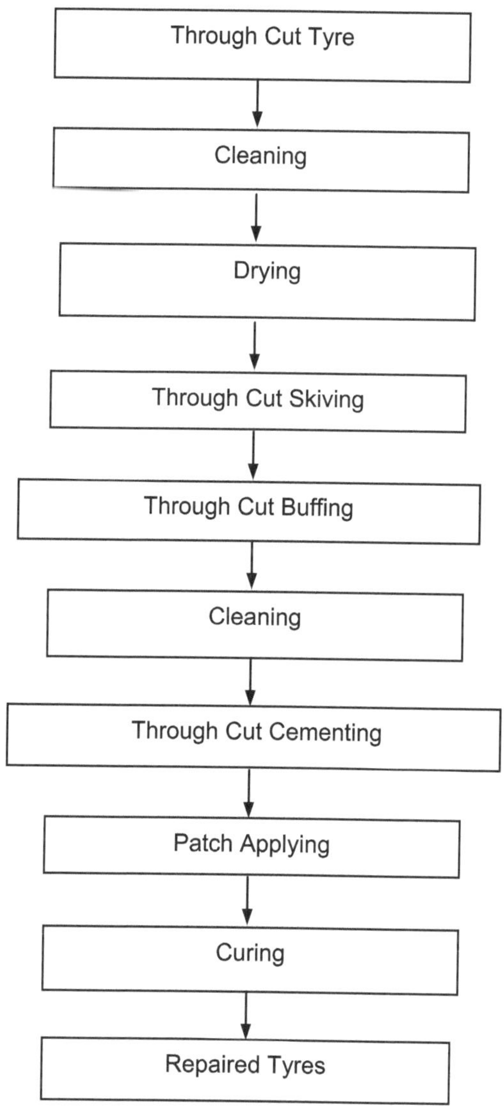

Fig. 2: Flow Diagram for Through Cut Repair

(v) *Cleaning:* The inner and the outer surfaces at the cut portion are cleaned by using compressed air, brush and petrol.

(vi) *Through Cut Cementing:* In this operation, the retreading cement is applied in two coats on the inner as well as on the outer surface of the cut area with brush and it is kept in this condition for a period of at least two hours to get dried.

(vii) *Patch Applying:* The cord fabric uncured patches are applied on the inner surface of cut portion and the outer portion is filled with cushion gum by using stitcher. Size of the patches would be according to the size of cut and the ply rating of tyre.

(viii) *Curing:* The curing of inner patches and the filled outer portion is done in sectional mould by passing the steam from a boiler at a pressure 3.5–5 kg/cm^2 and at a temperature of 120°–150° C for a period of 6 hours.

Now-a-days, precure patches are available in the market for a particular size of tyre injury and ply rating of tyre. These patches are applied from inside surface of the cut area. Localized curing of the repaired portion is done by the electric spotter. It has been found that the performance and life of tyres repaired with the use of precure patches would be better than the sectional mould repair as done conventionally.

9. Conclusions

The safe performance of a tyre is influenced by tyre design, tread pattern, rubber composition, inflation pressure and the conditions prevailing at the running area. A regular visual inspection and maintenance would pay the dividend by enhancing the tyre life. While in service only one fifth of the tyres is worn out, it gives a sense of economy to retread the tyre for multiple use. For a regular monitoring of the tyre performance, a performance recording system and a failure analysis system should be established for tyres. The corrective actions should be based on the outcome of the above systems.

10. References

[1] Matthan, R.K. 'Rubber Engineering'.

[2] Verma, M. 'Construction Equipment-Planning and Application'.

[3] Kumar, Prabhat "Design, construction and maintenance of haul road", presented in the Seminar "Equipment Maintenance and Productivity Drive", 2001-02, Central Coalfields Ltd.

[4] Darad, R.K., Ghosh, R. and Mukherjee, K. 'Maintenance efforts for OTR tyre life improvement with MS Access—A case study from Rajrappa Opencast Project' presented in the Seminar 'Maintenance and Productivity Drive', 2002-03, Central Coalfields Ltd.

Maintenance Efforts for OTR Tyre Life Improvement—A Case Study from Opencast Mines

R.K. Darad, R. Ghosh and K. Mukherjee

Excavation Section
Rajrappa Project, Central Coalfields Ltd.
Hazaribagh-829150

1. Brief About Tyres

Tyres are the black round rubberized components on dumpers, trucks and other mobile machines. In fact, these are much more sophisticated items than one might think. However, there is a science to the black art of making tyres properly and there are many critical factors that should be taken into account in the manufacture and operation of tyres. The complicated process of tyre manufacturing starts with raw rubber supplied in bailed form that is combined with some agent namely carbon black to modify hardness and strength of rubber. Oil-based softeners are also added to the rubber compound together with some chemicals to protect the finished product from the environmental hazards in which it would work. These softeners also allow the process of "curing in" during which the chemical linkages are developed between the long chain rubber molecules to give the compound its elastomeric properties. Such mixing process is the key to ultimate performance of the rubber compound and the tyre as a whole. With the advancement in technology, it has now been possible to improve the quality standards considerably. At the same time, the tyres have also been able to carry heavier loads, although this has not been simply a case of making the products of even larger diameters. However, there comes a point beyond which the tyres would become too bulky to move around. The generation of low profile ultra-heavy tyres are now changing the face of surface mining activities as these offer an improved economy, particularly in the expensive process of overburden removal and ore or coal transportation.

1.1 *Some Aspects of Tyres*

- Tyre cost is about 1/3rd of the total high cycle cost of the machine.
- Tyres are consumed in large numbers.
- Tyres are the capital intensive spares.
- Tyres are the components that are subjected to the ills of air pressure, indifferent loading pattern, poor operation, improper road or surface etc.

2. Tyre Specification and Standards

The tyres can be specified as below:

24.00	R	49	E 4	2-A
Tyre size 24.00 is the section width	Type–Radial	Rim Diameter 49 inches	TRA code	Specification Index, e.g. Cut Protected

The commonly accepted international standards are TRA (Tire and Rim Association, USA), ETRTO (European Tire and Rim Technical Organisation, Europe) and JATMA (Japan Automobile Tyre Manufacturers Association, Japan). The load capacity, inflation pressure, dimensions of the tire e.g. overall tyre diameter and tyre width, rim, tube, valves etc. are made as per the above standards.

3. Where the Rubber Meets the Road

The two important aspects in a mine system are (a) proper haul road design and its regular maintenance (b) inflation pressure of tyre, vehicle speed and load. Such factors would have a significant effect on the operational cost in mines. A good tyre and haul road combination would result in reduced fuel consumption, higher dumper speed, longer tyre life, more operational comfort, higher and safe working condition. The characteristics of good haul road are: good surface, adequate drainage, good sight distance to ensure safety, ample passing width, longest possible adverse grade, long curvature radius and super elevation at curves. Four simple factors should be kept in view to have a fair assessment of the haul road condition e.g. 1) smooth and clean road surface, 2) insufficient road width to allow dumpers to pass without driving on the berm 3) draining of water properly on the road surface, 4) proper operating speed of the dumper.

3.1 *Effect of Poor Haul Road Condition*

(a) *High Gradient:* The elimination of gradient on the haul road is the most common method to enhance tyre life. However, such concept is often not possible in a real life situation. The best alternative is to restrict the grade in the range of 5% to 8% on the rise and to minimize the length of grade. "The rule of thumb is that there is one-to-one ratio between the weight-transfer and the grade". However, on a 10% grade, the rear wheels would carry 10% more load than that they would bear on a level road.

A typical weight distribution upto 150 T capacity dumper would be as below:

A 8% grade would increase 8% weight on the Rear tyres (Up-grade loaded) and the 8% grade would increase 12% weight on the Front tyres

(Down grade/Empty). This may result in severe tyre overloading and may lead to structural or cut damage accompanied by much reduced value of fuel economy.

(b) *Improper Drainage:* A good drainage system for the haul road is essential for several reasons. Accumulated water on the road surface reduces the tyre traction and can conceal the tyre hazards e.g. rocks or ruts. Water also greatly increases the susceptibility of the rubber to cuts developed from sharp objects and can cause structural failures if the water seeps into the internal plies of tyres through these cuts.

(c) *Dumper Speed:* The surface material of the haul road determines the rolling resistance and traction coefficient, which in turn affects the speed of dumper and the gradability. It has been observed that the travel speed of dumper should be around 20–30 km/hr to achieve the optimum performance of the dumper. If the dumper speed is lesser than the above, it would result in a higher fuel consumption and slower rate of production.

It is generally recommended that for:

The maximum speed for Earthmovers should be	~ 64 kmph
The maximum speed for Graders should be	~ 40 kmph
The minimum speed for Dozer/Loader should be	~ 8 kmph

It has been observed that in mines, the dumpers are generally running at a speed of approx. 7–12 kmph.

4. Tyre Maintenance and Storage

While running the mines for 24 hours a day and seven days a week, it would go on adding up the costs involved in maintenance and replacement of tyres. Routine maintenance of tyres would reduce the vehicle downtime, reduce major repairs, improve operating efficiency and promote higher productivity. In order to achieve the maximum output from a dumper, it would be essential to maintain the tyres properly. An effective maintenance of the tyres can be done by following the simple steps like:

(a) A visual inspection of tyres has to be conducted before the start of dumper operation. It is required to look for the signs of irregular wear on the tread or shoulder of the tyre and examine the tyre for bubbles or bumps caused by air inflation or foreign objects. It should be repaired immediately as both such factors may lead to tyre failure and develop a dangerous working environment.

(b) The inflation pressure is another important factor. The air pressure should be checked daily for the dumpers that are in constant use because the air pressure is a critical factor in relation to the performance of tyres.

However, the air pressure may be checked weekly for the dumpers that have lesser demanding schedule.

(c) After fitting the tyres to a vehicle, it should be tested by operating the vehicle in full speed. Also the brakes should be applied in a sudden manner many a times and finally a check should be made for nut tightness as well as for the other fitments. This helps in avoiding wheel-stud failure and the problem of nut loosening.

(d) Dumpers with flat tyres, damaged or distorted rims or wheels, missing bolts or cracked studs should be taken out of operation.

(e) A deep crack, cut or other major problem may be noticed during the inspection of tyre and the dumper should be taken out of operation for necessary repair work under the circumstances.

(f) Care should be taken that no heat is applied or welding is done to join the parts of the wheel rim near the tyre. The application of heat is likely to cause serious damage to tyres and can cause their explosion. Tyres should always be removed before such jobs are carried out.

(g) Any mixing of tyres on dumpers should be avoided i.e. a tyre of normal tread depth should not be paired with that having a deep tread depth or a bias-ply tyre should not be paired with a radial tyre. Using two different types of tyres can cause damage to the other components of the dumper as the tyres would not work together to provide the desired traction effect and the expected performance in load handling would not be achieved.

(h) It is necessary to avoid the lifting of tyres through the center with a crane hook because this can damage the critical bead area. Instead, the tyre should be lifted under the tread by using flat straps. The flat straps are recommended over the wire rope or chain slings as these straps will not cause the cut or abrasion.

(i) The tyres should be stored properly when not in use. Also, the tyres should be placed in a cool, dry spot, away from direct sunlight to avoid premature aging. The tyres should be stored in standing upright position on the tread and the tendency to stack the tyres should be avoided; otherwise this may tend to weaken the tyres at the bottom of stack.

Tyre maintenance should be done regularly, as it would influence the entire site to keep the fleet of dumpers operating at the maximum efficiency. By following these simple steps, the mining operations can be done with a reduced investment on the tyres and boost the productivity level maintaining a safer environment.

5. Tyre Wear and Causes

The commonly observed modes of tyre wear and the causes of such wear are listed in Table 1.

Table 1: Modes of Tyre Wear and the Cause of Such Wear

Wear Modes	*Causes of Wear*
Rapid wear of Shoulders (Outer edges)	Under inflation, lack of tyre rotation, high speed applied in cornering
Rapid wear of Center	Over inflation, lack of tyre rotation
Cracked Treads	Alternating under/over inflation, high speed in driving, over loading
Wear on Sides	Lack of tyre rotation, defects in steering or suspension system
Feathered edge	Incorrect toe
Bald spots	Unbalanced wheel

6. Field Observations

The site conditions, driving skill for dumper operators, maintenance practices etc. should be duly considered while assessing the services given by tyres. Most of these factors are controllable and can be managed effectively to enhance the tyre life. It is necessary to highlight the following aspects in this connection:

1. The zone where mostly the damage due to cut occurs is not generally on the haul roads, rather these occur more at the dump areas, shovel faces etc., where restricted space often results in sharp turnout and in negotiating steep gradient. This is really very harmful for tyres. It has been seen that the rear tyre wears out more quickly than the front tyres because the operators spin the tyres while trying to get away from the pot-holes on the road. This also results in high incidence of spin cuts that can allow water to get into the steel cords and cause structural failure. This is also another reason as to why the dump areas need to be kept free from rocks and debris.

   Adequate turning radius of the dumpers can reduce the stresses on tyre sidewalls. This occurs while negotiating a sharp turning within a limited space. Hence, the tyres tend to get distorted when subjected to the powerful side-loading forces that occur during sharp turnouts. Inevitably, by maintaining a clean, unramped dump or loading area would help to reduce the frequency of sidewall cuts and failures.

2. "In a pneumatic tyre, the air contained carries the load and not the tyre casing". Hence, the key factor in tyre operation is the air pressure. Overloading or underloading of tyres reduces the tyre life and increases the operating cost. Even if the dumper tyres are kept properly inflated, it would still cut the costs. One of the most important factors in tyre maintenance is the attention to proper tyre inflation. Underinflated tyres loose their load-carrying capacity e.g. in a 24.00 R49 tyre, a 5 psi

reduction in air pressure equals a loss of 1130 lb of load-capacity/tyre. If a tyre is overinflated by 20%, it can loose as much as 30% of its expected service life. Hence, the tyre inflation should be checked on a regular basis, preferably daily and readings should be recorded on tyre record sheets to develop the maintenance history that would help to pinpoint the wear and failure trend of the tyres.

3. The tyre cuts would have a major impact on the cost. There are two issues that should be considered in this connection. The nos. of tyre that came out of service much early due to cuts (i.e. not delivering the potential life they could) and the potential hours lost due to the cut (i.e. how early in their service life were they cut). During the last 4 years, there were cut failures in 156 cases, retreading in 124 cases, burst in 86 cases and worn out ply in only 11 tyres out of total 384 failed tyres. About 41% of tyre cuts have been observed and these have been the highest amongst other defects. In short, a focus to decrease the cuts would perhaps lead to an increase in tyre life. The tyres opened for retreading have been the next figure as 33%. The burst cases were about 23% and only 3% cases were for the ply worn out tyres.
4. Normally during the warmer months of the year, the tyre performance becomes worse. This is because of the reason that the hotter the tyre runs, it affects the general performance of tyres badly. A hot tyre is more prone to rock cuts and would tend to wear faster. Further, the rainy season followed by the warmer months leads to possibility of more tyre failures as shown in Table 2.

Table 2: Possibility of More Tyre Failure during Rainy Season

Period of Climatic Change	*(2001–2002)*		*(2002–2003)*	
	B/D Frequency and %	*B/D Hrs and %*	*B/D Freq. and %*	*B/D Hrs. and %*
APR to SEP (Hot and Rainy season)	243 63%	11458 55%	98 68%	4159 79%
OCT to MAR	145 37%	9563 45%	46 32%	1080 21%

B/D–Breakdown

7. Application of MS Access

The data generated through the use of MS Access would help to go for future maintenance and repair plan of tyres. Table 3 denotes the summary of "*tyre failure*" in different years as per the data obtained from MS Excel. The compilation of the above data helps in understanding the trend of failure.

Table 3: Defectwise Tyre Failure

Reason for tyre removal (defects)	*Parameters*	*1999*	*2000*	*2001*	*2002*
Burst	Frequency	19	14	35	18
	Average Life (Hrs)	*4260*	*3380*	*3450*	*4218*
	Minimum Life (Hrs)	2525	2344	955	1882
	Maximum Life (Hrs)	6695	5549	7754	6815
	% of Total	*27%*	*13%*	*32%*	*27%*
Cut	Frequency	44	49	38	25
	Average Life (Hrs)	*2550*	*2870*	*2940*	*3560*
	Minimum Life (Hrs)	536	913	115	845
	Maximum Life (Hrs)	6323	5401	7537	7259
	% of Total	*37%*	*38%*	*30%*	*31%*
Retreading	Frequency	26	44	31	23
	Average Life (Hrs)	*4060*	*4050*	*4110*	*4838*
	Minimum Life (Hrs)	1288	2803	1609	2632
	Maximum Life (Hrs)	7000	6720	6725	7167
	% of Total	*35%*	*48%*	*34%*	*39%*
Ply worn out	Frequency	1	2	6	2
	Average Life (Hrs)	*1853*	*2032*	*3190*	*4197*
	Minimum Life (Hrs)	1853	1729	1486	3103
	Maximum Life (Hrs)	1853	2335	6465	5291
	% of Total	*0.5%*	*1%*	*5%*	*3%*
Frequency of Failure (Nos.)		*90*	*109*	*110*	*68*
Overall Average Life (Hrs)		*3341*	*3400*	*3446*	*4185*

8. Performance Evaluation of New Tyres

(a) As presented in Table 3, the overall frequency of tyre failure has reduced to a low value of 68 only in 2002 having a high average life during the past 4 years. The frequency of all types of failure has also reduced. The data from the table can be analysed further to note that the frequency of tyres opened for retreading is the main reason for tyre removal in the year 2000, 2001 and 2002. Also, there has been improvement in average, minimum and maximum life of tyres that were opened for retreading in 2002 i.e. as compared to those during the previous years.

The above data can be summarised further w.r.t. tyre removal as presented in Table 4.

Table 4: Data Summarised w.r.t. Tyre Removal

Ranking	*1999*	*2000*	*2001*	*2002**
	Reason for Tyre Removal			
1ST	Cut 37%	Retreading 48%	Retreading 34%	Retreading 39%
2ND	Retreading 35%	Cut 38%	Burst 32%	Cut 31%
3RD	Burst 27%	Burst 13%	Cut 30%	Burst 27%
4TH	Worn Out Ply 0.5%	Worn Out Ply 1%	Worn Out Ply 5%	Worn Out Ply 3%

In fact, the new tyre is removed before the tread depth of tyre wears upto 15 mm so as to utilize the remaining useful life after retreading. As a result, the percentage of Cut failure in terms of breakdown frequency is about 1/3rd of the total failure. The cases of tyre bursting are lesser. It is noted that the minimum tyre life has increased in all the cases w.r.t. those of last year as shown in Table 3.

(b) Another table generated by the MS Access may now be examined as presented in Table 5, to analyse the performance of tyres of different make i.e. "*make and defectwise tyre failure*". The yearwise data of tyre removal for retreading operation is shown in Table 5.

Table 5: Performance of Tyres of Different Make (Removed for Retreading)

Make	*Type*	*Parameters*	*1999*	*2000*	*2001*	*2002*
Bridge-stone (BS)	Radial	Frequency	13	5	1	-
		Average Life (Hrs)	*4865*	*4952*	*5053*	-
	Bias	Frequency	7	22	18	12
		Average Life (Hrs)	*2868*	*3599*	*3564*	*5220*
Good Year (GY)	Radial	Frequency	-	-	-	-
		Average Life (Hrs)	-	-	-	-
	Bias	Frequency	4	2	-	5
		Average Life (Hrs)	*3987*	*4748*	-	*3073*
Michelin (MH)	Radial	Frequency	-	2	-	-
		Average Life (Hrs)	-	*4818*	-	-
	Bias	Frequency	-	-	-	-
		Average Life (Hrs)	-	-	-	-
MRF	Bias	Frequency	-	-	-	-
		Average Life (Hrs)	-	-	-	-
Yoko-hama (YH)	Radial	Frequency	-	-	-	-
		Average Life (Hrs)	-	-	-	-
	Bias	Frequency	2	13	11	6
		Average Life (Hrs)	*3188*	*4257*	*4968*	*5546*
Total Frequency of Failure (Nos)			26	44	30	23
Overall Average Life Before Retreading (Hrs)			*4063*	*4055*	*4128*	*4838*

It is observed from Table 5 that the lowest average life i.e. 3073 hrs. for Goodyear make bias tyres has been recorded as in the year 2002. Although the same make of tyre has given better life to the tune of 3987 hrs and 4748 hrs in the year 1999 and 2000 respectively, it may be mentioned that the lower performance of tyre may be due to change in tread design. The reason for the same could be that the rubber content of the tyres developed later was more than that provided in previous tyres e.g. the tread depth of earlier tyres was 63 mm, while the latest tyre is having 70 mm tread depth.

The Bias type tyre of Bridgestone make has a reduced life as compared to the tyres of Yokohama make. Moreover, the frequency of failure for Bias type Bridgestone tyre is higher than that of Yokohama make during the last 4 years.

The first life of the Yokohama tyres is better than that of the Bridgestone make tyres. Table 5 indicates that the overall performance of tyres of different make should be examined.

(c) Table 3 has been obtained from the MS Access i.e. failure of tyres of different manufacturers. The yearwise data for tyres has been compiled as shown in Table 6.

Table 6: Failure of Tyres of Different Make

Make	*Parameters*	*1999*	*2000*	*2001*	*2002*
Bridge-stone (BS)	Frequency	55	67	58	31
	Average Life (Hrs)	*3210*	*3150*	*3470*	*4655*
	Minimum Life (Hrs)	536	913	769	1865
	Maximum Life (Hrs)	7000	6720	7754	7259
	% of Total	*59%*	*57%*	*53%*	*46%*
Good Year (GY)	Frequency	11	2	2	18
	Average Life (Hrs)	*4026*	*4748*	*1930*	*2858*
	Minimum Life (Hrs)	2890	4327	1064	1882
	Maximum Life (Hrs)	5374	5169	2793	5291
	% of Total	*15%*	*3%*	*1%*	*26%*
Michelin (MH)	Frequency	8	3	-	-
	Average Life (Hrs)	*5360*	*4142*	-	-
	Minimum Life (Hrs)	4229	2791	-	-
	Maximum Life (Hrs)	5695	5170	-	-
	% of Total	*14%*	*3%*	-	-
MRF	Frequency	-	-	-	*1*
	Average Life (Hrs)	-	-	-	*2149*
	Minimum Life (Hrs)	-	-	-	*2149*
	Maximum Life (Hrs)	-	-	-	*2149*
	% of Total	-	-	-	*2 %*

(Table 6: Continued)

(Table 6: Continued)

Yoko-hama (YH)	Frequency	16	37	50	18
	Average Life (Hrs)	*2320*	*3706*	*3480*	*4817*
	Minimum Life (Hrs)	609	959	115	845
	Maximum Life (Hrs)	5259	5146	7537	6944
	% of Total	*12%*	*37%*	*46%*	*26%*
Total Frequency of Failure (Nos)		*90*	*109*	*110*	*68*
Average Life of Tyres (Hrs)		*3341*	*3400*	*3446*	*4185*

Table 7 depicts the trend in performance of tyres make-wise as per data obtained from Table 6.

Table 7: Rank of Failure of Tyres Makewise

Rank of Failure	*1999*	*2000*	*2001*	*2002*
	Make and Failure (%)			
1st	BS-59 %	BS-57 %	BS-53 %	BS-46%
2nd	GY-15 %	YH-37 %	YH-46 %	GY&YH-26%
3rd	MH-14 %	MH & GY-4 %	GY-28 %	
4th	YH-12 %	GY-3 %		MRF-2 %

Table 7 shows that the failure of Bridgestone make tyres is the highest during the last 4 years and contributes more than 46% of failure. The Yokohama make tyres occupy the next position and contributes to more than 26% of failure. Although the frequency of removal of the Goodyear tyres due to various defects is much lesser (ref. Table 6), the average life and the maximum life during the year 2001 and 2002 are observed to be the lowest.

Thus, it is seen that with the help of MS Access many valuable tables can be generated that would help in the performance evaluation of tyres and the analysis of defects would be easier and more informative.

(d) The trend in tyre life due to various defects can be noted as presented in Table 8.

It is seen that there is major improvement in case of ply wornout for tyres as its 1st average life is having an increasing trend and the improvement in tyre life during the year 2002 is 24% as compared to that in the year 2001. However, for the other defects namely burst, cut and retreading, the average life improvement have been 18%, 17% & 15% respectively. Moreover, the life improvement has been observed in case of all the defects in the year 2002 as compared to those in the year 2001.

Table 8: Trend in Tyre Life Defectwise

Year	*Type of Defect*	*Frequency*	*Average Life (hrs)*	*Minimum Life (hrs)*	*Maximum Life (hrs)*
1999	Burst	19	4260	2525	6695
2000		14	3380	2344	5549
2001		35	3450	955	7754
2002		18	4218	1882	6855
1999	Cut	44	2550	536	6323
2000		48	2870	913	5401
2001		38	2940	115	7537
2002		25	3560	845	7259
1999	Retrea-ding	26	4060	1288	7000
2000		44	4050	2803	6720
2001		31	4110	1609	6725
2002		23	4838	5005	7167
1999	Ply Worn Out	1	1853	1853	1853
2000		2	2032	1729	2335
2001		6	3190	1486	6465
2002		2	4197	3103	5291

(e) Similarly, the tyre failure data has been compiled tyre typewise for different years as shown in Table 9.

It is evident from the above data that the radial tyres are performing better than the Bias tyres and that the minimum life and maximum life of the radial tyres are more than those of the Bias tyres. The frequency of failure for Bias tyres is much larger than that of Radial tyres. The average life of Bias tyres is 4200 hrs. in 2002 which is comparable to that for Radial tyres in 2002, although the same was lesser in 1999, 2000 and 2001. The population ratio of Bias Vs. Radial tyres is approx. 90%–100%. Hence, the chance of failure of Bias tyres is also more.

Table 9: Performance of Different Types of Tyres Yearwise

Type	*Parameters*	*1999*	*2000*	*2001*	*2002*
Bias	Frequency	59	100	108	63
	Average Life (Hrs)	2570	3290	3410	4200
	Minimum Life (Hrs)	536	913	115	845
	Maximum Life (Hrs)	5374	6146	7754	7167
	% Failure	50%	89%	97%	93%
Radial	Frequency	31	9	3	5
	Average Life (Hrs)	4810	4610	5480	3990
	Minimum Life (Hrs)	1855	2791	5053	2600
	Maximum Life (Hrs)	7000	6720	5906	7260
	% Failure	50%	11%	3%	7%

9. Payoffs: Breakdown Reduction

The breakdown hours and the frequency of failure from April to March in 2002 year comes to be lesser by about 75% and 63% respectively as compared to those of the previous year. Moreover, the contribution of tyre failure in 2002 is only 5% of the total no. of failures for 85 T dumpers as shown in Table 10:

A comparative study on the breakdown hours of tyres during 2001–02 and 2002–03 has been presented in Table 10. The failure of tyres could be controlled to some extent after Dec 2001. However, the breakdown of tyres has gone on the increase from Jun to Sep in 2002 as apprehended earlier and discussed above in Sl. 5 (IV). Further, the downtime for tyres is considerably lesser as compared to the previous year.

Table 10: Breakdown Hours and Frequency of Failure of Tyres

Year	*2001–2002*		*2002–2003*		*Change in B/D Hrs w.r.t. Earlier Year (%)*	*Change in Failure Freq. w.r.t. Earlier Year (%)*
Month	*Break-down Hr*	*Frequ-ency*	*B/D Hr*	*FREQ*		
Apr	1204	72	176	16	(–) 85%	(–) 78%
May	1108	37	167	11	(–) 85%	(–) 69%
Jun	1363	23	857	17	(–) 37%	(–) 26%
Jul	1947	37	1108	23	(–) 43%	(–) 38%
Aug	2735	37	807	14	(–) 70%	(–) 62%
Sep	3101	38	1044	19	(–) 66%	(–) 55%
Oct	2891	37	231	7	(–) 92%	(–) 81%
Nov	1801	24	320	10	(–) 82%	(–) 58%
Dec	2006	21	256	11	(–) 87%	(–) 48%
Jan	1470	23	132	8	(–) 91%	(–) 65%
Feb	937	19	34	3	(–) 96%	(–) 84%
Mar	458	21	107	7	(–) 77%	(–) 67%
Cumulative B/D hr and freq of failure	21021	388	5239	144	(–) 75%	(–) 63%
Total Breakdown hours of Dumper	141804	4833	99390	1497		
Tyre Breakdown (%)	15%	8%	5%	10%	(–) 64%	20%

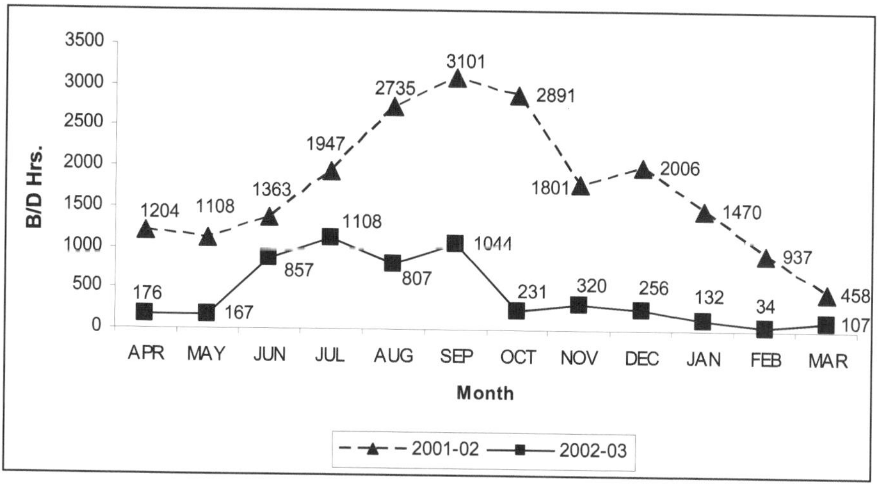

Fig. 1: Comparison of Tyre B.D. Hrs. for 85 T Dumpers at Rajrappa

The total breakdown hours on account of tyre failure was 21,021 hrs. in 2001 and the same has reduced to 5,239 hrs. in 2002. This shows that there has been a drastic reduction of 15,782 breakdown hours i.e. a reduction of 75% as compared to that during the previous year. Although the frequency of tyre failure in 2002 has increased from 8 to 10% as compared to that during the previous year, the tyre breakdown hours has reduced from 15% to 5% in 2002 w.r.t. that in 2001.

10. Conclusions

A proper care and strategic actions i.e. regular visual inspection of tyres, tightness of fixing studs and nuts and inflation pressure has paid the dividend during the year 2002. It has been observed that a drastic reduction of 15782 breakdown hours has taken place in 2002 as compared to that during the previous year over the same duration i.e. the breakdown on account of tyre failure has reduced by 75%. In case of frequency of failure, there has been a reduction of 244 nos. (i.e. 63% lesser) have occurred in 2002 for tyres. The 41% cut failure (highest) and 23% for burst failure (2nd highest) would bring in the net result i.e. 64% of downtime for tyres during the last 4 years. These two defects leading to tyre failure and the low average life of tyres indicate that differential tyre inflation, improper haul road condition and slow dumper speed are the major contributing factors. The focus has to be converged now to carry out extensive checking and maintenance of the proper tyre pressure and proper condition of the haul road. The information generated from MS Access has been useful to understand the failure related statistics defect-wise, make-wise and tyre type-wise. The feedback obtained from the performance data as above plays a dominant role in strategic planning and managing the maintenance of tyres.

Selection and Maintenance of OTR Tyres—Failure Analysis and Remedial Measures

N. Thangavelu

Donimalai Iron Ore Mines
National Mineral Development Corporation Ltd., Donimalai-583118

1. Function and Construction of OTR Tyre

The tyre alongwith its rim fitment makes a kind of pressure vessel capable of supporting the load at a predetermined degree of deformation. The deflection or the degree of deformation of tyre complements the vehicle suspension. The OTR tyres are designed for a higher deflection as compared to the highway tyres. These tyres are fitted in earthmoving equipment to work under widely diverse conditions. The OTR tyres are characterized by very high load bearing capacity and these are able to negotiate abnormally rough terrain. The requirements of OTR tyre design are vehicle cushioning, traction, overcoming the effects of excessive loads, high speeds and heat. Hence, each component of the OTR tyre e.g. bead, casing, sidewall and tread are of heavy-duty construction.

2. Constructional Features of Cross-Ply (Bias-Ply) Tyre

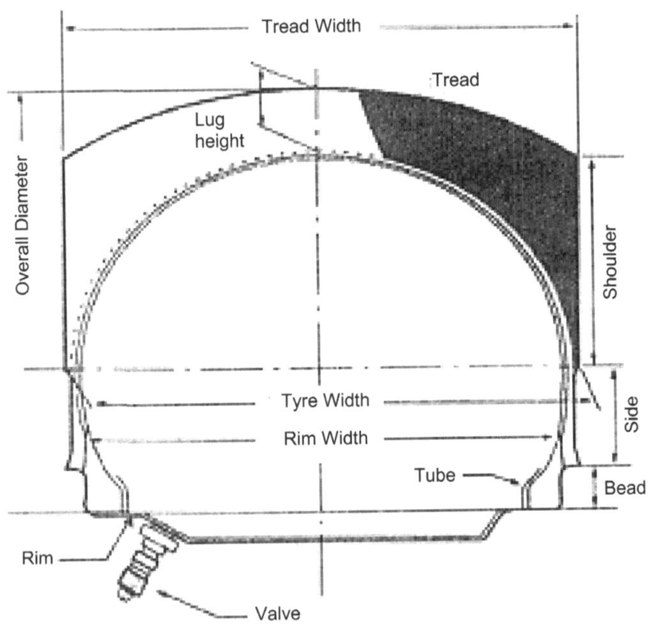

Fig. 1: Components of tyre

2.1 *Tubed Tyre*

(a) *Beads:* The tyre beads anchor the tyre to the rim and may be considered as the foundation of tyre. All plies are tied-in to these bundles of strong wire, which prevent any change of shape.

(b) *Sidewalls:* The protective rubber cover on the outer sides of casing are called sidewalls. These are designed to flex and bend without developing any crack, when subjected to ordinary continuous deflection or any sudden shock of very high degree.

(c) *Casing:* The layers of rubber-cushioned cord or fabric make up the casing. The strength built into this part of the tyre confines the inflation pressure that supports the load and cushions the shocks developed.

(d) *Plies:* The plies are the successive layers of casing. Each cord in the ply is completely surrounded by resilient rubber compound. Also each ply is insulated from the next ply. The term "ply rating" as defined by the Tyre Industry, is an index of tyre strength and does not necessarily indicate the number of cord plies in the tyre.

(e) *Breakers:* The breakers are the extra plies located under the tread. These are meant to absorb any sudden shock and provide an extra resistance to the tyre against the blows and impacts received from the rough surfaces or from obstacles.

(f) *Tread:* The tread is the part of tyre that comes in contact with the road. It must provide for traction, resistance to wear and cuts.

(g) *Tube and Flap:* The function of the tube is to retain air or inert gases under pressure within the casing. The flap protects the tube from any injury that can be caused by the rim and tyre beads.

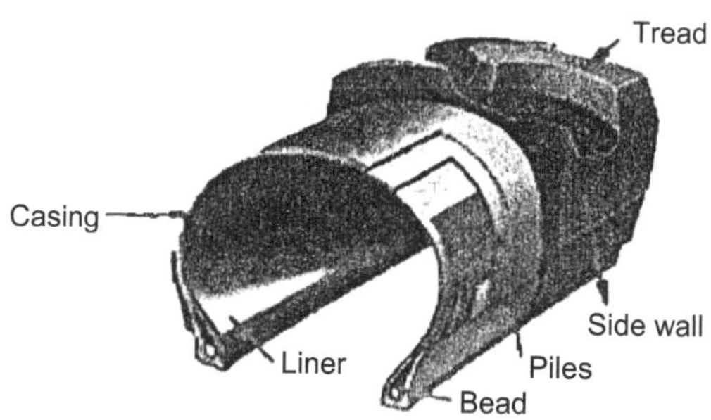

Fig. 2: Construction of Bias-ply Tyre

2.2 *Tubeless Tyre*

Apparently, this resembles a tubed tyre. However, there is no separate tube in such design. In tubeless tyres, the inside of tyre has an airtight lining extending beneath the bead to effect a complete air seal with the rim when inflated. A flap is not required with a tubeless tyre.

2.3 Radial Tyre

The casing has only one radial ply and the crown is stabilized by a belt made of several plies (steel plies).

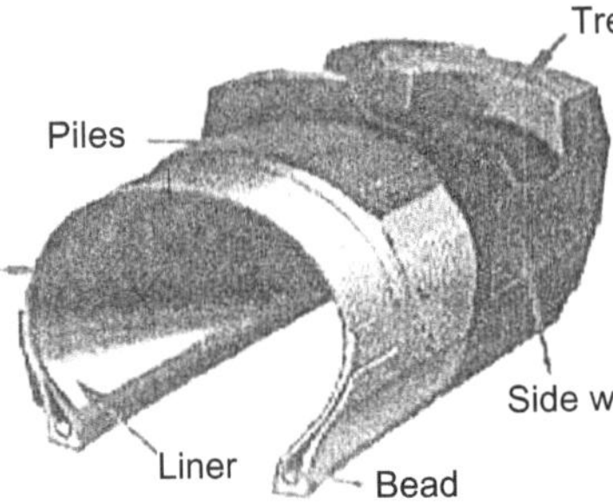

Fig. 3: Constructional features of Radial Tyre

3. Selection of Tyres

Definition of the terms used in the selection of tyres for Heavy Earth Moving (HEM) Equipment:

a) Size and Ply Rating

Example: 21.00 × 35, 36 PR

21.00 represents the nominal cross sectional width of the tyre in inches. 35 is the nominal diameter of the tyre from bead to bead in inches and it indicates the size of the rim diameter. The ply rating is an index of tyre strength and it does not necessarily mean the number of plies actually used as the base material in tyre.

b) Standard and Deep Tread Tyres

The standard tread depth of tyre 18.00 × 25, 32PR comes to be 32 mm; whereas the deep tread of the same tyre is 54 mm.

c) Tubed and Tubeless Tyres

The tubeless tyres offer some important merits over the tubed tyres as given below:

1. Tyre downtime gets reduced due to the elimination of tube and flap troubles.
2. It is possible to have quicker and easier fitting.
3. It is easier to handle.
4. Tyre and flap replacement costs are eliminated.
5. Cooler running—It is particularly important for high ambient temperature and in the negotiation of longer haul lengths.
6. The possibility of bursts get reduced. Un-noticed accidental damage can lead to a burst with tubed tyres. Such damage shows as a slow

leakage for tubeless tyres. It is safer to have repairs before more costly damages are developed.

d) Bias-Ply and Radial Tyres

The merits of Radial tyres over Bias-Ply tyres are:

1. Longer tyre life
2. Excellent grip
3. Provides excellent flotation
4. Smoother ride because of the increased vertical flexibility.

e) Several Types of Construction of OTR Tyres

Type-A = Highly resistant to cutting, hacking and abrasion.

Type-B = Specially designed to handle heat build-up on long runs and intense service conditions.

Type-C = For high speed travel on long hauls.

f) Index to Tyre Classification

The tyre industries all over the world have adopted a uniform code which indicates the amount of rubber in the tread, type of tread pattern and its application.

Code Number	*Tread Pattern Type*	*Type of Service*
E-1	Rib	Haulage
E-2	Traction (regular)	Haulage
E-3	Rock (regular)	Haulage
E-4	Rock deep tread	Haulage
E-7	Flotation	Haulage
G-1	Rib	Grader
G-2	Traction (regular)	Grader
G-3	Rock (regular)	Grader
G-4	Rock deep tread	Grader
L-2	Traction (regular)	Loader, Dozer
L-3	Rock (regular)	Loader, Dozer
L-4	Rock deep tread	Loader, Dozer
L-5	Rock extra-deep tread	Loader, Dozer
L-3S	Smooth	Loader, Dozer
L-4S	Smooth deep tread	Loader, Dozer
L-5S	Smooth extra-deep tread	Loader, Dozer

g) Factors related to Jobsite and Vehicle

i) Vehicle

- Type of equipment (Dumper, Loader, Grader).
- Laden axle weight, unladen axle weight.
- Tyre size of the original equipment.

ii) Site condition

- Type of material being loaded.
- Condition of haul road surface and soil condition of haul road (includes consideration of the weather and its effects on the haul road e.g. mud, snow as well as haul road turning radius, super elevation etc.).
- Condition of loading and dumping areas.

iii) Use of vehicle at site

- Working-day-averaged speed (Average Number of miles or kilometers covered in one hour). Such average speed is considered based on total travel of the vehicle during the whole day. In this connection, it is necessary to have a record of the following:
  1. Length of cycle (round trip).
  2. Number of cycles covered in a day.
  3. Number of working hours in a day.
- Ton-Mile-Per-Hour ratings (TMPH) and Ton-Kilometer-Per-Hour ratings (TKPH).
- Maximum loaded speed.
- Proper application of the machine.

iv) Other considerations

- Combined behaviour of vehicle and tyre (in traction, flotation etc.)
- Previous tyre performance, including:
  (a) How wear takes place on the tyres.
  (b) Reasons for final removal of the tyres.

h) Calculation of TKPH of OTR Tyre

$$\text{TKPH} = \left(\frac{Q_L + Q_E}{2}\right)\frac{NL}{H}$$

Q_L = Tyre load (in Tonne)—Laden Mass of Vehicle.
Q_E = Tyre load (in Tonne)—Empty mass of vehicle.
N = No. of Cycles covered in a day.
L = Cycle Length (round trip) in kilometre.
H = No. of working hours in a day.

Example:

TKPH for 50 Ton Dumpers (H-50, BEML Dumper) for Donimalai Iron Ore Mine (for the rear axle).

$$\text{TKPH} = \left(\frac{13+10}{2}\right)x\frac{75x5}{18} = 218 \text{ TKPH}$$

4. Maintenance of Tyres

4.1 *Proper Inflation*

A proper inflation pressure is essentially required to achieve the best performance of tyres. Indeed, it would be possible to obtain optimum traction, flotation and load endurance when proper inflation pressure is maintained. Both overinflation and underinflation would shorten the tyre life and can result in tyre failure. The proper inflation pressure would depend on the technical features of the vehicle, ground conditions, load, speed etc.

The correct inflation pressure should be determined on the basis of maximum load on each tyre and average speed of the vehicle. Considering all such factors, the inflation pressure is to be selected from the table/chart recommended by the manufacturers.

4.1.1 *Effect of Underinflation*

- Excessive deflection developes generating heat and this leads to early tyre failure.
- Tread and ply separation.
- Cord fatigue gets accelerated leading to the development of broken cord.
- Sidewalls become susceptible to rupture.
- Tread wear becomes uneven and radial cracks develop.
- Rim becomes displaced creating air leakage in tubeless tyres.
- Cracks develop at the inner liner.

4.1.2 *Effect of Overinflation*

- Increased ground contact pressure at the center of tread developing excessive wear at the middle portion of tread.
- Reduced protection of the cord against shocks from uneven road surfaces resulting in vulnerability to rupture from cuts or shock.
- Excessive pressure is exerted on the beads increasing the potential for the beads to burst.
- Riding comfort deteriorates and tendency to slip develops.

The inflation pressure should be checked for the OTR tyres in hot and cold condition. Also the inflation pressure should be checked for the OTR tyres in cold condition at least once in a week. Excessive pressure buildup due to increase in temperature during operation of the vehicle should never be reduced by “bleeding”. The excessive pressure may be due to increase in temperature of tyre while in operation. In case excessive pressure is noticed in all the tyres, load and speed should be reduced to bring the tyre temperature to the normal operating temperature. Figure 4 presents the trend in variation of tyre life (%) due to the change in inflation pressure (%) under the conditions of proper load and appropriate inflation pressure (100 psi).

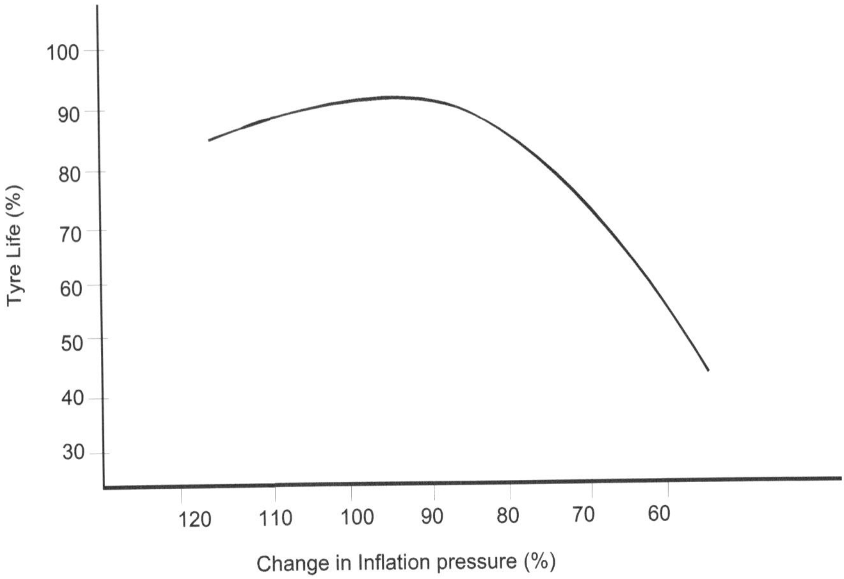

Fig. 4: Inflation Pressure and Tyre Life

A tyre inflation checklist duly signed by the technician and the Shift In-charge should be made available datewise to record the inflation pressures of different tyres. The recommended inflation pressure values for the front and rear tyres of 50 Ton Dumpers are given below:

- Front Tyre : 80 psi (Bias-Ply) and 110 psi (Radial)
- Rear Tyre : 85 psi (Bias-Ply) and 115 psi (Radial)

If the inflation pressure is lesser in any of the tyres, the same should be recorded and corrected to the recommended inflation pressure.

4.2 *Load*

An overloading of tyre would reduce the tyre life and increase the chance of an early tyre failure. For the best tyre performance, the maximum recommended load should not be exceeded. If the load exceeds the specified capacity of tyre, a tyre of higher ply rating should be used.

Results of Overloading

- Excessive heat generation causes ply separation.
- Excessive tyre deflection causes broken cords.
- Excessive tread movement against the road surface develops rapid wear.
- Excessive bead movement causes bead failure.
- Risk of bursting of tyre due to increased cord tension.

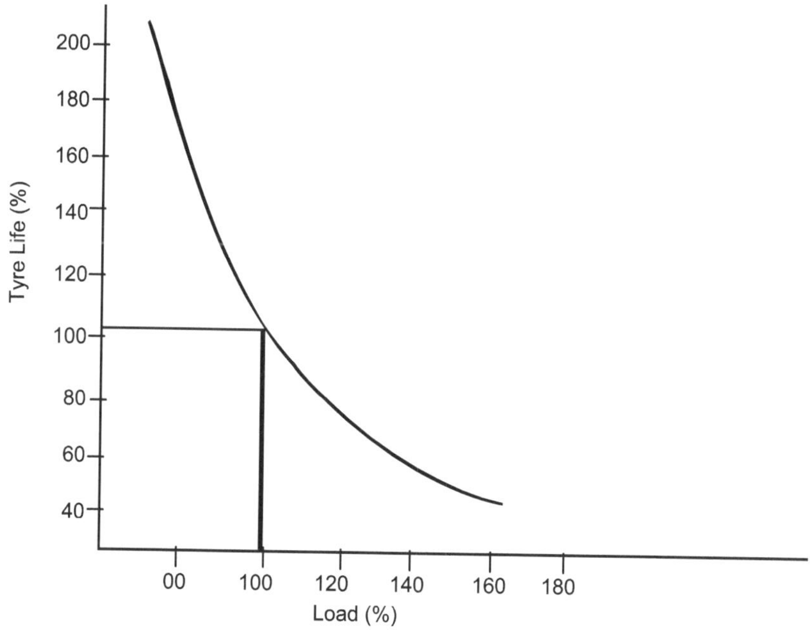

Fig. 5: Load Vs Tyre Life

4.3 *Speed*

An excessive travel speed of the dumper produces abnormally high internal temperature in tyres. A vehicle has two speed limitations i.e. the actual maximum speed that the vehicle can attain and the average operating speed that the vehicle sustains. The average sustainable operating speed to be limited for the tyres as Ton-Mile-Per-Hour (TMPH) rating.

4.4 *Ample Tyre-To-Vehicle Clearance*

The maintenance staff should carefully check all the tyres mounted on each vehicle to make sure that the tyres do not rub against any part of the vehicle, either on straight drive or at the turnings. Failure to ensure an ample tyre-to-vehicle clearance would result in premature tyre replacement. The common sources of such problem are Spring clips, U-bolts, Fender bolts etc.

4.5 *Wedged Stones*

A proper maintenance is required for overcoming the problem of stones or other objects that have become wedged between dual tyres or between the tyres and the vehicle. In such situation, the stone should be removed promptly to prevent any serious tyre damage. A permanent rock ejector should be installed on the vehicle, if possible. However, the condition of the rock ejector should be checked periodically and any wornout rock ejector to be replaced in due time. A worn-out rock ejector will lead to the damage of tyres.

4.6 *Matching of Dual Tyres*

The maintenance staff should be specially careful in mounting or servicing the dual tyres to make sure that the tyres have the desired diameters within the given tolerance. Even the new tyres of different make or a newly retreaded tyre and a new tyre may not match on a dual assembly. A new tyre should never be mounted along side a tyre worn out considerably as it would lead to the difference in diameters. Further, the radial tyre should not be matched with bias-ply tyres.

4.7 *Prompt Repair of Tyres*

A prompt maintenance inspection is required for detection of cuts, snags or other damages to enable timely repair in a proper manner and to prevent the possibility of tyre blow out. Water, dirt or other materials are likely to enter the cuts or other kinds of tyre damage may take place that might get extended upto the casing. Eventually, the tread or ply separation may occur and it would lead to a blow out. It is very important that even damages that appear to be minor apparently should be repaired properly at the earliest.

4.8 *Keeping Tyres Free From Oil, Grease and Gasoline*

Rubber absorbs oil, grease or gasoline quickly. After absorbing such materials the tyre swells and becomes soft and spongy. Any problem of this kind would lead to a permanent damage and it may result in fatal accident. The tyres should never be cleaned with gasoline or other petroleum products. Also the tyres should not be allowed to stand in puddles of petroleum products (or at the areas saturated with petroleum products). If any petroleum product is getting impregnated on the tyre, the tyre should be promptly flushed off or wiped off with high pressure water jet.

4.9 *Misalignment of Axles*

Misalignment is likely to cause abnormal and rapid wear of the tread of tyres. Usually excessive toe-in or toe-out is the reason for such problem. Misalignment usually produces a "feather edge" on the tyre shoulder because the tyre is being dragged sideways as it starts revolving and the tread is being scuffed off constantly in one direction.

4.10 *Tyre Removal from Equipment*

Sometimes, the tyres are removed early for retreading; even if some useful tread depth is remaining. The left-out tread depth can give further life to the tyre. Also, the tyres are removed for retreading after complete tread wear. In such case, the treader may not accept the tyre for retreading as inner ply separation might have taken place in the tyre. The tyres should be taken away from the equipment for the purpose of retreading as per recommend-dations of the manufacturer for tread wear.

4.11 *Handling of Tyres*

Improper handling of tyre can lead to its damage, especially to the beads. Hence, it is necessary to prevent any excess pressure from being exerted on the beads.

4.12 *Mounting and Dismounting of OTR Tyres*

Before demounting the tyre, all air should be exhausted from a single tyre and from both the tyres of a dual tyre assembly. Also this should be done prior to removing any other component of the wheel e.g. nuts, rim clamps, etc. A broken part of the rim can get blown apart under pressure and may cause serious injury or death of the operating personnel at the nearby area.

During mounting of the tyre, it should not be inflated before all the components are placed properly. The tyre should be placed in a safety cage or using any other restraining device. It should be inflated to a pressure of approximately 5 psi and the components should be rechecked for a proper assembly. It should be checked that the O-ring does not roll out of its groove. If the assembly is not proper, the tyre should be deflated and necessary corrective measures should be taken. An inflated or partially inflated tyre/rim assembly should never be hammered. If the tyre assembly has been done in a proper manner at a pressure rating of 5 psi (approx), it should be made to inflate to have a perfect seating on the beads.

4.13 *Road Surface Maintenance*

The maintenance of road surface is one of the most important factors in determining the life of a tyre. Bumps, chuck holes, rocks etc. would tend to develop cut and wear of the tyres. Even tyre bursting may occur due to such problems.

4.14 *Tyre Storage*

In general, the tyres should be removed from the rim, cleaned and stored in a cool and dry room. It should be stored in an upright position (not horizontally) so that a proper space between the beads can be maintained. It should be ensured that the tyre does not get deformed by any external pressure exerted on it. The following should be avoided when storing the tyres:

- Direct sunlight
- Ozone
- Oil and grease
- High temperature and humidity.

5. Failure Analysis of OTR Tyres

Some studies have been done on the failure of 21.00 × 35 OTR tyres used for 50 T Dumpers at the Donimalai Iron Ore Mine. The project has discarded 33 Nos. of 21.00 × 35 E-4 OTR Tyres of Bridgestone, Dunlop, Goodyear and Yokohama make and 20 Nos. of Radial tyres of Michelin and Toyo make during the period April 2001 to August 2002. The details are given in Table 1.

Table 1: Different Makes of Radial Tyres

Make	*Bias Ply/Radial*	*Qty*	*Average Life in Hrs. (total)*	*No. of Tyres Failed within 3000 Hrs.*
Bridgestone	Bias Ply	14	4237	06
Yokohama	Bias Ply	12	4066	02
Goodyear	Bias Ply	05	4319	01
Dunlop	Bias Ply	02	1791	02
Toyo	Radial	03	4905	Nil
Michelin	Radial	17	6009	02

The failure details of the tyres which had life less than 3000 hours are given in Table 2.

Table 2: Failure Details of Tyres

Sl. No.	*Tyre No.*	*Make*	*New/ Retreaded Tyre (RT)*	*Life in hrs. (Actual)*	*Reasons for Failure*
1.	F9U381327	Bridgestone	New	1546	Sidewall bulged
2.	F1S386176	Bridgestone	New	2397	Inner ply separation
3.	F1S386447	Bridgestone	New	2246	Sidewall bulging
4.	F1R387230	Bridgestone	New	2089	Inner ply separation
5.	S2J003123	Bridgestone	RT	2843	Sidewall burst
6.	F0R387616	Bridgestone	New	2741	Inner ply separation
7.	XN7ECB535	Yokohama	RT	2362	Tread burst
8.	AN3ECB447	Yokohama	New	2895	Sidewall burst
9.	D459704	Goodyear	New	2773	Inner ply separation
10.	A066MAR91	Dunlop	RT	1182	Sidewall burst
11.	A680JUL90	Dunlop	RT	2400	Inner ply separation
12.	DPZ0119L3A	Michelin	New	2630	Bulging on the tread
13.	XPB0013V5A	Michelin	New	2345	Tread separation

5.1 *Reasons for Tyre Failure*

A. Sidewall Bulging

- Overload
- Running continuously in underinflation
- Manufacturing defect (heavy air pockets formed at the sidewall).

B. Inner Ply Separation

- Running continuously in underinflation
- Overload
- Manufacturing defect (loose ply).

C. Tread Burst

- Overload
- Overspeed
- Underinflation
- Overinflation
- Mismatching of tyre
- Impact of rock

D. Tread Separation

- Overload
- Overspeed
- Underinflation
- Overinflation
- Manufacturing defect (like loose bonding of tread with crown).

The following measures as presented in Table 3 should be taken to avoid the failures of OTR tyres:

Table 3: Recommended Measures to Avoid OTR Tyre Failure

Sl. No.	*Reasons of Failure*	*Measures Recommended*
1.	Overinflation of the tyre	Ensure recommended inflation pressure and check for correct tyre gauge.
2.	Underinflation	Ensure proper inflation pressure. Also check for any slow leakages.
3.	Toe-in or toe-out becomes incorrect	Check for wheel alignment.
4.	Faulty brake operation	Check brake drum overheating and condition of the service brakes on all the wheels and for the effective braking.
5.	Tread cuts	Tread cuts to be repaired promptly. Avoid sharp objects and stones on the road.
6.	Tread chipping out	The dumpers should be driven carefully. Avoid sharp objects during vehicle operation.

(Table Contd.)

(Table Contd.)

7.	Through cut	Through cuts due to sharp edges of the boulders or stones weakens the cord of the tyre ply. This reduces the resistance of tyre body.
8.	Dual wall touching	This is possible only due to overloaded tyre or underinflation of the tyre. Due care should be taken by the operators in this respect.
9.	Scoring	Foreign objects trapped between the sidewalls of the rear tyre. This should be immediately attended.
10.	Sharp metal objects protruding outside	Metallic objects like bolts, rod pieces, other metallic pieces of mild steel etc. should be removed from the haul roads.
11.	Cutting due to rubbing	Similar action may be taken as above.
12.	Accidental damage on side wall	While in operation dumper operator should take care to avoid stones, boulders etc.
13.	Wear due to overload	The supervisor/engineer should check for proper loading and inflation of tyre and for the correct quantum of load to be carried by the dumper.
14.	Shoulder fatigue due to overload	Same as above.
15.	Brake drum over-heated and overloaded	Check for proper brake linings and for proper load on the dumpers to avoid such damages.
16.	Tyre bead damage due to improper use of mounting tool	Proper mounting tool should be used as per recommendation of Earth Moving Equipment Manufacturers or Maintenance Manual for tyres provided by tyre manufacturers.
17.	Damage due to improper mounting	Avoid unskilled labour in tyre maintenance. Also avoid careless mounting or dismounting of tyres.
18.	Damage due to ring flange	The damaged ring flange may cause damage to the tyres on uneven surface of the road. Hence, the ring flange in such conditions should be discarded whenever necessary.

6. Conclusions/Recommendations

- Proper OTR Tyre should be selected for use on the vehicle.
- It is necessary to follow proper maintenance practices.
- It is essential to take care of the haulage roadway condition.
- Overload and overspeed of the vehicle should be avoided.
- The tyre should be removed timely from the vehicle for retreading/repairing.
- It should be tried to enhance the life of tyres.
- The tyre cost should be reduced.

Tyres in Mining Industry—Repair and Maintenance

J.N. Singh, R.L. Rai, S.N. Singh, A.K. Pandey, Sandeep Shukla and R.N. Tripathi

Kusmunda Arca, South Eastern Coalfields Ltd.
Seepat Road, Bilaspur-495001

1. Introduction

A glance into the story of wheels would reveal that the wheel mechanism plays an important role in the working of mobile machinery. Proverbially, the wheel is used as the "wheel of fortune" going round and round. The wheel was invented by the Persians in a perfect way around 500 BC. It is said that around 2500 BC i.e. during the Indus Valley Civilisation, some improvement on the design of wheel was made. Thereafter, two wheels were joined with a shaft for use in vehicles. In order to reduce the vibration, a rubber lining of circular shape was placed on the wheel rim and this helps to increase the speed of the mobile vehicle. Subsequently, the tyres with tubes for general applications and tubeless tyres for heavy earthmovers were introduced to enable higher running speed as well as for prolonging the life of tyres.

Thus, the quest of mankind first realized the need for wheels that was an amazing discovery of the ancient time. Later the inquisitiveness of human being led to invention of tyres to be mounted onto the wheel rim. This was followed by many trials to finally manufacture small tyres for light and medium size vehicles and finally the heavy duty tyres for use in Heavy Earth Moving Machineries (HEMMs) i.e. 35T, 50T, 85T, 120T and 170T dumpers, Water Tankers, Graders, Cranes, Wheel-mounted Dozers, LHDs etc. deployed in mining projects. Basically, there are two types of tyres:

(a) *Cross Ply:* These are generally called as nylon tyres having lesser life and lesser toughness due to increased wear and tear as compared to radial tyres.

(b) *Radial Tyres:* These are the tyres that are made by the blending of nylon with woven wire for higher life, more toughness and better appearance.

Table 1: HEMM and Size of Tyres

Sl. No.	*Type of HEMM*	*Size of Tyres*	*Remarks*
1.	35 T	18 × 25	
2.	50 T	21 × 25	
3.	85 T	24 × 49	
4.	120 T	30 × 51	Tread width 30 inch and inner diameter of tyre 51 inch.

Table 2: Working Life of Tyres

Sl. No.	*Type of HEMM*	*Size of Tyres*	*Working Life in Hrs.*
1.	35 T	18 × 25	5000 (Normal Life) 2000–after 1st retreading 1500–after 2nd retreading.
2.	50 T	21 × 35	6000 (Normal Life) 2000–after 1st retreading 1500–after 2nd retreading.
3.	85 T	24 × 49	10000 (Av. Life) 2500–3000–after 1st retreading. 1500–2000–after 2nd retreading.
4.	120 T	30 × 51	11000 to 12000 (Av. Life) 3000–after 1st retreading. 2000–after 2nd retreading.

2. Life of Tyre

The life of tyre largely depends on soil condition as well as the design of haul road. The harder the soil/haul road surface, the lesser would be the life of tyre. The haul road designs should be scientific and it should be made of tar (Tar Haul road).

3. Different Make of Tyres

The Original Equipment Manufacturers (OEM) of tyres throughout the world including those in India are mostly multinational companies. The tyres of different manufacturers as used in opencast mines are given as below:

1. Bridgestone
2. Yokohama
3. Goodyear
4. Michillene
5. Toyo
6. Modi
7. Apollo
8. Vikrant
9. M.R.F.
10. J.K.

The cost of maintenance and repair of tyres for retreading is given in Table 3.

Table 3: Tyre Maintenance and Repair Cost

Sl. No	*Type of Tyre*	*Retreading Cost*	*Tread Width (mm)*	*Tread Depth (mm)*	*Overall Diameter (mm)*	*Cost of New Tyre*
1.	30 × 51	Rs. 77000/-	765	80	2895	Rs. 257103/-
2.	24 × 49	Rs. 36500/-	588	75	2500	Rs. 68181/-
3.	21 × 25	Rs. 17000/-	520	55	2000	Rs. 62000/-
4.	18 × 25	Rs. 7400/-	460	45	1620	Rs. 64849/-
5.	35/65-33 (24 PR Type)					Rs. 195750/-

4. Inflation of Tyres

There are three types of tyre inflation

- Overinflation
- Underinflation
- Normal inflation.

a. Overinflation and Causes

- Driver fatigue
- Rapid centre wear
- Reduced bruise resistance
- Reduced traction
- Wear and tear of suspension components.

b. Underinflation and Causes

- Ply separation
- Tread separation
- Tread rubber reversion
- Circumferential breaks
- Upper sidewall cracking
- Bead area separation
- Loose inner liners
- Loose inside cords
- Run flat
- Uneven wear
- Fast shoulder wear
- Heel and toe wear-lug tyres.

c. Normal Inflation and Performance of Tyre

- Long life
- Better comfort for operator
- Least wear and tear

- Increased vehicle speed
- Balanced drive.

5. Tread Design Aspects

(a) Smooth tyre
(b) Rib type tyre
(c) Lug type tyre
(d) Hybrid type tyre.

a. Smooth Design

Merits

- Maximum wear resistance
- Good cut resistance
- Maximum traction on hard surface.

Demerits

- More heat buildup
- Lack of penetration/traction on soft ground.

b. Rib Type Design

Rib type design of tyre is most suitable for the operation of HEMMs. It has the following merits:

1. Maximum lateral traction
2. No heat problem
3. Excellent steering response.

c. Lug Type Design

These tyres have the following merits in operation.

1. Maximum directional traction
2. Maximum breaking force
3. Excellent for intermittent service.

d. Hybrid Design

It is the blend of smooth, ribbed and lugged type of tyre design. This design has the following merits.

1. Such tread design is much used in earthmoving applications.
2. It is a compromise design providing the desired properties as obtained from the three basic tread pattern designs i.e. smooth-ribbed-lugged.

6. Industry Codes

The codes of Goodyear tyres are given below:

- E1-100 LEVEL RIB
- E2-100 LEVEL TRACTION
- E3-100 LEVEL ROCK

- E4-150 LEVEL ROCK
- E7-100 LEVEL FLOTATION

6.1 *Dozer and Loader*

- L2-100 LEVEL TRACTION
- L3-100LEVEL ROCK
- L4-150 LEVEL ROCK
- L4S-150 LEVEL SMOOTH
- L5-250 LEVEL ROCK
- L5S-250 LEVEL SMOOTH

6.2 *Grader*

- G1-100 LEVEL RIB
- G2-100 LEVEL TRACTION
- G3-100 LEVEL ROCK
- G4-150 LEVEL ROCK

7. Computerised Maintenance and Repair of Tyres

Routine condition monitoring or periodic maintenance and repair is done as per computerized schedule after every shift/week/month/six months/year, particularly in terms of the number of hours of operation of the HEMMs as well as taking due care of primary, secondary and tertiary operation of the earthmovers.

(i) ***Tube Tyres:*** Tube tyres with inflated tubes up to optimal pressure are generally used in light and medium vehicles. The vehicles operating with tube tyres are having safety tags attached on the vehicle. The overinflation of tyres may lead to bursting.

(ii) ***Tubeless Tyres:*** Tubeless tyres are usually used in heavy duty earth movers i.e. dumpers of capacity 120T, 170T and 220T for the transportation of overburden. These tyres provide safe operation of the vehicles with optimal air pressure.

8. Safety Aspects

(a) It is necessary to check the rims regularly and ensure that there are no fatigue cracks and/or weakened welds.

(b) Cracked or broken rims or rim components must always be replaced. Welding of rim or its components should never be tried.

(c) Very high temperature may develop inside the tyres (due to jammed brake) and this can cause explosion through self-ignition. The possibility of such danger can be avoided by inflating the tyres with nitrogen gas instead of compressed air.

(d) When inflating a tyre, it is recommended to keep a safe distance of 5 m.

(e) When dealing with twin wheels, it is necessary to deflate the inner tyre before undoing the outer wheel.

(f) Radial tyres should be properly lubricated i.e. from the bead-toe to the "GG-groove" in order to facilitate the mounting and ensure a correct seating.

Table 4: Types of Tyres and Service Condition

Vehicle	*Service Condition*	*Type of Tyre*	
		Bias	*Radial*
Truck	Long hauls/high speed	E3, Type 2S	E2/E3, Type 2S
Truck	Short hauls/poor-to-fair underfoot	E4, Type 4S	E4, Type 4S
Scraper	Soft underfoot/ medium hauls	E2, Type 4S	E2, Type 4S
Scraper	Mixed conditions	E3, Type 4S	E3, Type 4S
Loader	Dig and load/soft underfoot	L2, Type 4S	L2, Type 4S
Loader	Dig and load/soft rock	L5, Type 6J	L5, Type 6S
Grader	Mixed service	G2, Type 4S	G2, Type 4S

9. Summary

- Retreading of tyres should be done strictly as per recommended schedule and under the guidance of maintenance supervisor.
- First retreading is more economical.
- Second retreading is also economical, but upto a certain extent.
- The heavy earthmoving machinery tyres are mostly of tubeless type.
- In order to achieve a better life of tyres, the primary, secondary and tertiary haul roads should be of scientific design and should be repaired and maintained timely.
- Inflation of tyre should be very near to optimal inflation as there are chances of more wear and tear as well as discomfort to operator due to overinflation. Underinflation may lead to poor speed of vehicle, more wear and tear as well as reduced efficiency of operator.
- Optimal inflation of tyre can provide better efficiency and improved utilization of the vehicle. Further, there would be cost reduction in primary and secondary retreading of tyres in such approach.

10. Conclusions and Recommendations

- For heavy earthmoving machinery the tyre must be purchased only from the original equipment manufacturer having the proven track record for quality of product commensurate with the global technology level in tyre manufacturing.
- Tubeless tyre is the only option for HEMMs and ensure the equipment to be more efficient in terms of optimal utilization and cost effectiveness.
- It is also necessary to go for the authorized company having proven record so as to achieve the best performance of retreaded tyres.

This paper has been prepared based on the experience gained and practice of tyre repair and maintenance at the Excavation Workshop for HEMMs namely Workshop No. 1, Kusmunda Opencast Project (World Bank Aided) as per discussions held with the engineers engaged in the repair and maintenance including retreading of tyres for HEMMs. The computerized periodic maintenance practice followed at the Excavation Workshop No. 1, Kusmunda Opencast Project, SECL has also been kept in view.

Reference

[1] Goodyear tyre manual, 2002.

Off-Highway Tyres in Opencast Mines

Deepak Vidyarthi

Kudremukh Iron Ore Mines, Kudremukh Iron Ore Company Ltd.
Kudremukh-577142, Chikmagalur

1. Introduction

The word 'tyre' is synonymous with everyday's mobility in life. With the emergence of a large number of leading manufacturers of tyres, some familiar names of companies are available in mining industry like Bridgestone, Good Year, Toyo, Mitsui, Firestone, Dunlop etc. With the increasing trend in trackless mining specially in underground mines, the use of tyres is also on the increase for wheel mounted machinery like LHDs (Load-Haul-Dump), LPDTs (Low-Profile-Dump-Trucks), Scoop Trams, Portable Compressors etc.

However, the total volume of production from the underground mines as compared to that from the opencast mines would be hardly 20% in India. The opencast mines that are the major users of large HEMMs, account for a higher consumption of tyres. However, the tyre consumption has showed upward or downward trend over the past few decades. Such trend relates to speed, ease of overall maintenance and maneuverability of equipment. With the increase in size of heavy machinery, particularly for the dump trucks of capacity varying from 25 T to 370 T, the demand for large sized tyres has increased multifold. The tyre cost poses to be the major component in overall mining cost.

2. Types of Tyres

Two types of tyres are commonly used in opencast mines, i.e. the Bias tyres and the Radial tyres.

2.1 *Bias Tyres*

Bias tyres of any pattern consist of a carcass, tough bead wires and breakers.

2.2 *Radial Tyres*

These tyres comprise of a casing, tough bead wires and belts. The internal thickness of rubberized material depends on size, pattern and ply rating.

Fig. 1 depicts the structural diagram of Radial and Bias tyres.

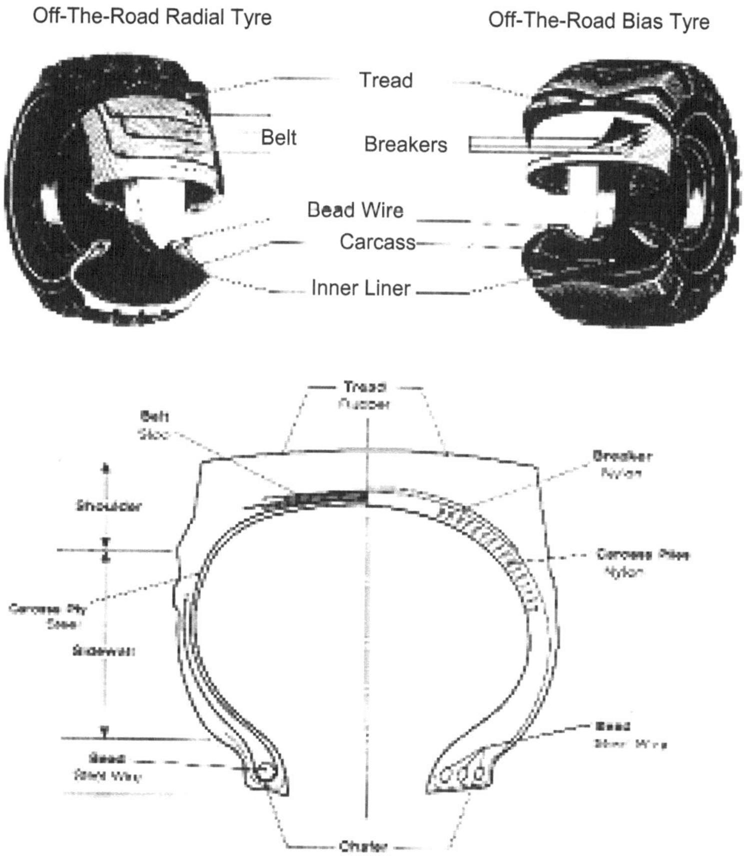

Fig. 1: Structural Diagram

2.3 *Structural Details*

While the construction of OTR tyres largely depends on the intended use, the following constructional features are common.

- Tread
- Carcass
- Beads
- Breakers
- Side Walls
- Inner Liner.

2.3.1 *Tread*

It is the outermost covering of the tyre and is the only part that normally comes in contact with the road surface. Depending on the intended use of tyre, the rubber compound applied to the tread is modified to customize cut resistance, heat resistance and oil resistance. The tread pattern also influences the performance of tyres. Various tread patterns are available like Rock pattern, Rock-flush pattern, Rib pattern, Traction pattern, Block pattern etc.

2.3.2 *Carcass*

The carcass consists of a number of rubber coated layers of fabric called "plies". It forms a semi-rigid framework for compressed air but it is flexible enough to absorb some shock and jolts.

2.3.3 *Beads*

The beads fix the tyre to the rim to support the load.

2.3.4 *Breakers*

The breakers are rubber coated layers of cord between the tread and the carcass to bind them together.

The breakers prevent cuts in the tread from reaching the carcass and absorb the shocks also.

2.3.5 *Side Walls*

The side walls consist of a flexible crack-resistant rubber to protect the carcass from possible damages.

2.3.6 *Inner Liner*

The inner walls of tubeless tyres are lined. The liner is made of an air-impermeable rubber compound and is comparable to tubes used in the tube-type tyres.

3. Selection Criteria

The following factors should be considered in the selection of Tyres for HEMMs keeping in view the high initial cost of tyres:

- Mobility
- Traction
- Abrasion resistance
- Cut and impact resistance
- Wear resistance
- Carrying capacity
- Tyre availability

4. Tyre Life

The life of tyres and the consequent unit cost (Rs/ton handled) depend largely on (a) haul road condition, (b) average haul distance, (c) proper and timely inflation of tyres maintaining the minimum recommended air pressure. Some statistics is given in Table 1 as obtained from the Kudremukh Iron Ore Mines. Kudremukh Iron Ore Company deploys 120 T Dump Trucks using 30 × 51 size of tyres with 52 ply rating to meet an annual target of hauling 20.0 million tones of highly abrasive iron ore. The fleet size of the Rear Dump Trucks at Kudremukh Mines is 32 that comprises of 25 WABCO, 5 DRESSER and 2 UNIT RIG Trucks. The average life of tyres at KIOCL during the past decade has been to the tune of 4060 hours against the ROM production that varies from 16.0–20.0 million tons per annum.

The factors affecting the life of tyres at Kudremukh Mines are:

(a) Severe monsoon season with an average annual rainfall of 7000 mm over a period of four months (June–September). This renders the site condition quite arduous and has a direct bearing on the life of tyres.

(b) Hardness and abrasive nature of the ore.

(c) Long haul distances giving rise to high temperature of the tyre.

5. Types of Failures

The common failures for OTR tyres at Kudremukh Mine are given below, indicating the percentage of failure during the last five years:

Table 1: Types of Tyre Failures at Kiocl

Sl. No	*Type of Tyre Failure*	*Year*				
		1998–1999	*1999–2000*	*2000–2001*	*2001–2002*	*2002–2003*
1.	Centre Wear	4%	6%	7%	4%	1.9%
2.	Normal Wear	4%	0%	9.9%	8.9%	15.8%
3.	Cut on Tread	46%	61%	40.8%	46%	59.5%
4.	Side Wall Cut	30%	31%	40.8%	40.3%	22.2%
5.	Cut on Shoulder	9%	0%	1.5%	0.8%	0.6%
6.	Misc. Failures Including Bulge	7%	2%	0%	0%	0%
		100%	100%	100%	100%	100%

6. Safety Aspects

The earth moving tyres are packed with compressed air having more than 100 psi air pressure. These are the sources of potential danger in the event of accidental burst and can hurt a person severely. Hence, certain 'Safety Precautions' as listed below must be observed strictly to avoid any untoward incidence.

6.1 *To Do*

- Inflate tyres in a safety cage.
- Replace weak or damaged parts.
- Replace all severely rusted rims.
- Check for excessive side ring play and ring buttering.
- Double check the tyre and the rim before inflating.
- Always deflate tyres prior to demounting.
- Inspect wheel nuts and clamps periodically.

6.2 *Not to Do*

- Do not inflate a tyre outside a safety cage.
- Do not mix rim parts of different make or models.
- Do not overinflate tyres.
- Do not use damaged, rusted or distorted rims.
- Do not use oversize tyres without proper rims.
- Do not overload tyres or rims.
- Do not reinflate a tyre that has been run flat without inspecting the tyre, rim and wheel for any damage.

6.3 *Safety Warning*

In order to protect the life and limbs of the operator, all the 'mounting' and 'de-mounting' procedures and safety practices as recommended by the manufacturers should be understood and followed meticulously.

6.4 *Tyre Handler*

The tyre handlers are used at Kudremukh Mines for changing the tyres of 120 T trucks. One tyre handler of IMTCO Ltd and one of Komatsu Ltd are in use. The tyre handler is capable of pulling out tyres from the trucks and rotating them through 95° above horizontal and 360° continuous rotation in the lateral direction. The time taken for changing a tyre of 120 T dump truck using this machine would be about one hour as compared to that required in the manual changing of a tyre with the help of crane taking about six hours.

6.5 *Tyre Guards*

The tyre guards are made of steel chains and are commonly used with all sizes of HEMM tyres. Tyre guards are extremely helpful in protecting the tyres from wear and tear or direct damages and ensure a better life of the tyres (Fig. 2).

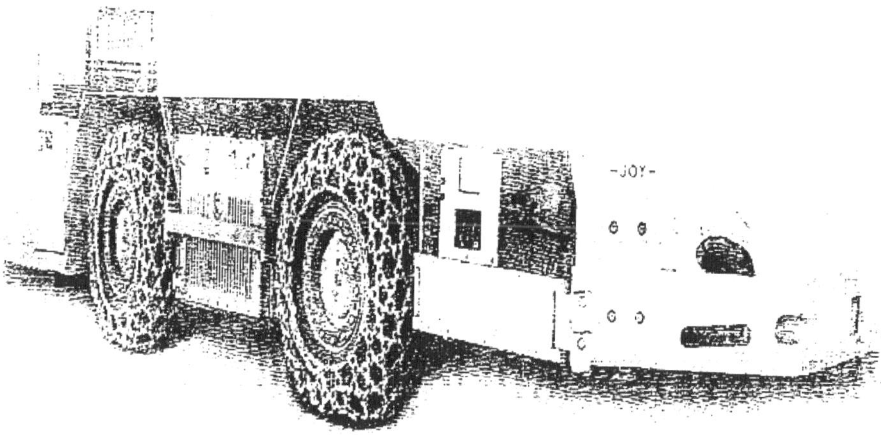

Fig. 2: Tyre Guards

7. Latest Technological Advancement

With the latest technological advancement, the Good Year & Co has brought in a new concept in tyre manufacturing through the introduction of "Rubber Twinset" or "Two-Piece Tyres". The Company has started manufacturing and field testing of two-piece tyres in the size range of 51″ (1.3 m) and 57″ (1.5 m). These tyres are suitable for Komatsu, Euclid, Unit Rig and Liebherr Trucks.

7.1 *Construction of Two Piece Tyres*

The two pieces of tyres consist of (i) Casing and (ii) Tread belt that fit together using interlocking grooves and are held together by air-pressure. High tensile steel cable is incorporated into the tread making the tyre slightly heavier than traditional tyres (about 500 lbs more) as in Fig. 3.

Fig. 3: Two–Piece Tyre

The merits in the use of such two piece tyres are:

- Damages to the tread will not affect the casing.
- The assembly offers flexibility in tread design.
- The tread belt package can be designed to suit the needs of changing conditions in a mine.
- The assembly can handle nicks and cuts in a better manner than the tyres used traditionally in mines.
- As the tread belt package protrudes beyond the side wall more than that in a traditional tyre, it provides additional protection for the casing also.
- The two piece tyre can cut 'down time', especially during the changing of tyres.
- This offers a better life and haulage speed combination.

8. Conclusions

- The OTR tyres are essentially for use in large opencast mines to handle the large volume of bulk solids.
- The tyres can be made specifically to meet the compounding needs of a mine commensurate with the latest technological advancement.
- In view of the spiraling cost of tyres, it becomes obligatory on the part of operators to take utmost care in operation and maintenance to avoid cut and impact injuries to tyres.
- In order to ensure a better life of tyres, the maintenance of haul roads of optimum width and to keep it clean has become essential accompanied by regular grading and sprinkling of water.

9. References

[1] Bristol Voss, Rubber Twin-Set, WME, Oct 2003.

[2] Deepak Vidyarthi, Mining Practices at Kudremukh Iron Ore Mine, MEAI-2001.

[3] Goutam Roy & S. Sengupta, Selection of Right Tyre for Opencast Mining, MEAI-2001.

[4] Bridgestone Manual.

[5] Yokohama Handbook.

Earthmover Tyres in Bailadila Iron Ore Project (Deposit No. 14 & 11C)

V.K. Jain

Deposit No. 14 & 11C, Bailadila Iron Ore Project,
National Mineral Development Corporation Ltd.
Kirandul-494556, Dantewara

1. Introduction

The main objective for increased use of earthmoving machinery is to achieve higher productivity and reduce the operational cost. Reduction in tyre consumption would play the most important role in reducing the cost of production.

Basically, a tyre with its recommended rim fitment forms a terodial pressure vessel to be capable of supporting the rated load at a predetermined degree of deformation. The deflection of the tyre complements the vehicle suspension.

The earthmover tyre has to work under diverse operating conditions, some of which are high load, abnormally rough terrain etc. Hence, for designing the earthmover tyre, the requirements of vehicle cushioning, traction, means of overcoming effects due to excessive loads, high speeds etc. are to be examined carefully. The issue gets further complicated with the consideration of operational requirements e.g. ruggedness, strength etc. All these are the essential requirements in earthmoving equipment operation. Each component of the tyre i.e. bead, cord body, sidewall and tread are of heavy duty construction to meet the requirements of rough and tough operating condition.

2. Tyre Components

A. Tubed Tyre

(a) Bead
(b) Sidewall
(c) Casing
(d) Plies
(e) Breakers
(f) Tread
(g) Tube and Flap

B. Tubeless Tyre

This resembles to the tubed tyres with no separate tube. However, the inside of the cover has an airtight lining that extends beneath the bead to effect a complete air seal with the rim when inflated. The flap is not required for the tubeless tyres.

The Different types of tyres used at the Bailadila Iron Ore project for the earthmoving equipment are given in Table 1.

Table 1: Different Types of Tyres Used for Earthmoving Equipment

Sl. No.	*Equip-ment*	*Capacity/ Model*	*Equip-ment Make*	*Tyre Size*	*Ply Rating*	*Maxi-mum Load (kg.)*	*Tyre Inflation (psi)*
1.	Dumpers	50 T	BEML/ CAT	21.00 × 35 Bias tyres (Tubeless)	36	11125	90
2.	Motor Grader	OD 805	BEML	1300 × 35 (Tubeless)	22	2530	30
3.	Dumpers	85 T	CAT	27.00 × 49			90
4.	Mobile Crane	8 T	Escorts	13.00 × 24	12	2485	35

The tyre consumption is much more in 50 T Dumpers as compared to the tyres used in other vehicles. Every year about 90 nos. of new tyres are used costing about Rs. 100 lakhs. The average life obtained for the new tyres from the manufacturers namely Bridgestone or Goodyear is only about 2300 hrs. of operation. Previously, the radial tyres were used in the project and each tyre was costing Rs. 2 lakhs. However, the use of radial tyres has now been stopped because no firm was available for the retreading of radial tyres and these tyres were used till destruction.

3. Parameters Influencing Tyre Life

(i) **Position of Tyre on Machine**

Tyre wear varies considerably depending on the location; the tyres are fitted to i.e.

- driving wheels twinned for Dumpers
- steering wheel for Dumpers.

(ii) **Inflation Process**

(a) It provides support against the load. It has to be measured and set in "cold" after the vehicle has been made to stand for a long time.

(b) By increasing the tyre flexing under inflation causes rise in internal temperature of the tyre.

(c) If the tyre is overinflated it will lack grip on the road and will be more susceptible to crown impact as tread hacking.

(iii) **Tyre Loading**

Each tyre size is having a load limit as recommended by the manufacturer.

(iv) Influence of driving factors should be duly considered like sudden braking, rate of acceleration, cornering at high speed, causing an increase in temperature.

(v) Configurations of the haulage roads and the quality of maintenance of the roads are of equal importance. The layout and transverse profile of the haulage road influences the dynamic overloads and cornering forces. High dynamic overload and lateral skidding would lead to a greater risk of tread separation. The maintenance of the roads greatly influences the likelihood of tyre damage at the side walls or at the treads.

(vi) The hours worked by the vehicle in a day is considered as an important parameter that plays a great role in heat generation.

(vii) Proper inflation of the tyre is essential for good tyre performance. The correctly inflated tyre permits all portions of the tread to have a contact with the road surface. Yet these tyres result in excessively high internal heat and premature failure. Overinflation develops excessive wear at the center of tread. The tyres and the related expenditure can be saved by applying the proper inflation pressure.

(viii) The tyre inflation should be checked after the vehicle has been kept idle for a sufficient period of time to allow the tyre to get cool and reach the ambient temperature. It is important that the air pressure is checked and corrected only when the tyre is at normal temperature. Also 24 hours of operation does not permit the checking of inflation pressure of the tyre, because the tyre is not completely cooled. For such cases, a correction factor can be determined by experiments. It is necessary to check as many tyres as possible when these are cold and again after at least two hours of operation. The average difference should be added to the recommended pressure when checking is done for the tyres during constant operation. This difference will get decreased when the tyre stands and gets cooled.

(ix) When the hot pressure exceeds the recommended cold pressure by more than 15 psi, the speed or the load must be reduced. The only way to reduce excess pressure is to reduce the speed or the load.

(x) The overloading makes the tyres to get damaged. It is very important to estimate the actual load and restrict the volume of materials to be carried; whenever it is necessary to ensure that the tyres are not overloaded. In order to obtain the maximum life of tyres, it is essential to operate the tyres within the load or inflation pressure.

4. Experience with Some New Tyres

The reasons for the replacement of new tyres of 21.00×35 size and 36 ply rating are:

- Sidewall bulging

- Cord lifting
- Ply separation

The causes for these problems are:

(a) Tyres running in low air pressure.

(b) Loss of inflation due to leakage through the tyre-rim interface. Constant running of the tyre under lower inflation would lead to excessive stress development at the flex point resulting in liner crack followed by cord lifting and subsequently a seepage. Seepage of air through the tyre-rim interface would lead to loss of inflation.

(c) Moderate-to-high loading and the constant running of tyres would lead to stress development at the flex point triggering the above action.

(d) Improper maintenance of mine roadway.

The remedial measures to be taken in this context are:

(a) Regular monitoring of the performance of tyres in operation.

(b) Thorough inspection and critical examination of the failed tyres would help to reduce further failure of tyres.

Study of Tyres in West Bokaro Mines—An Overview

Sureshwar Singh and P.N. Tiwary

West Bokaro Mines, Tata Iron and Steel Company Ltd.
Hazaribagh-825314

1. Introduction

The West Bokaro mines of Tata Steel are located in the district of Hazaribagh. The location of the colliery is at a distance of 36 km from the Hazaribagh town, 200 km NW of Jamshedpur and 71 km from Ranchi. The colliery has a lease hold area of 4300 acres and the total mineable coal reserves would be about 290 MT (Fig. 1).

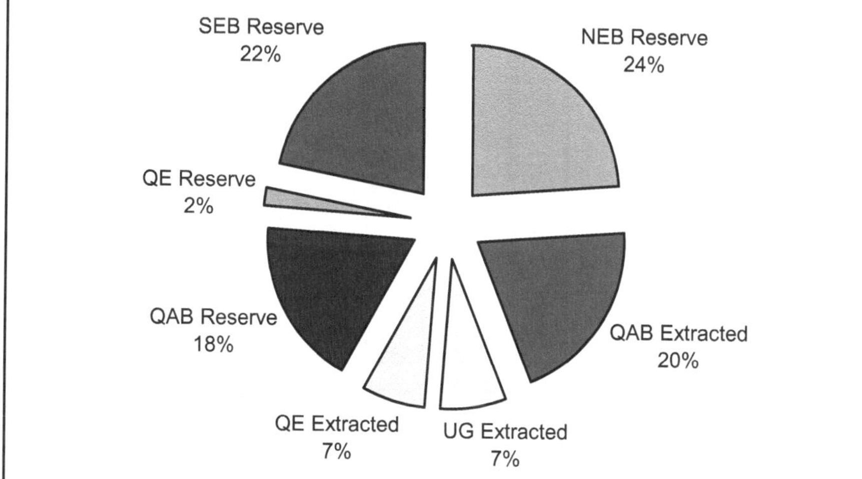

Fig. 1: West Bokaro Coal Reserves

The coal production at West Bokaro comes from the opencast mines. Beneficiation of raw coal is done at the washery to produce clean coal and middlings. The transportation of coal to the railway siding is done by aerial ropeways. The loading of coal is done at the railway siding for despatch to the Tata Steel plant located at Jamshedpur.

2. Tyre Mounted Equipment

A large fleet of tyre mounted vehicles are in operation at this project. The list of major tyre mounted equipment deployed at West Bokaro mines is given in Table 1.

Table 1: Major Tyre Mounted Equipment at West Bokaro Mines

Equipment	*Make*	*Model*	*Capacity*	*Tyre Size*
Rear Dumper	CAT	773B/D	50 T	21.00 × 35
Bottom Dumper	CIPL	HM-1160C	60 T	18.00 × 33
Pay Loader	CIPL	HM-2071	8 M^3	29.5 × 29
Water Sprinler	CIPL	HM-1035	28 KL	18.00 × 25
Motor Grader	BEML	BG-825A	-	18.00 × 25

The yearwise expenditure on tyres for use in earthmoving machinery deployed at the West Bokaro coal mines is given in Fig. 2.

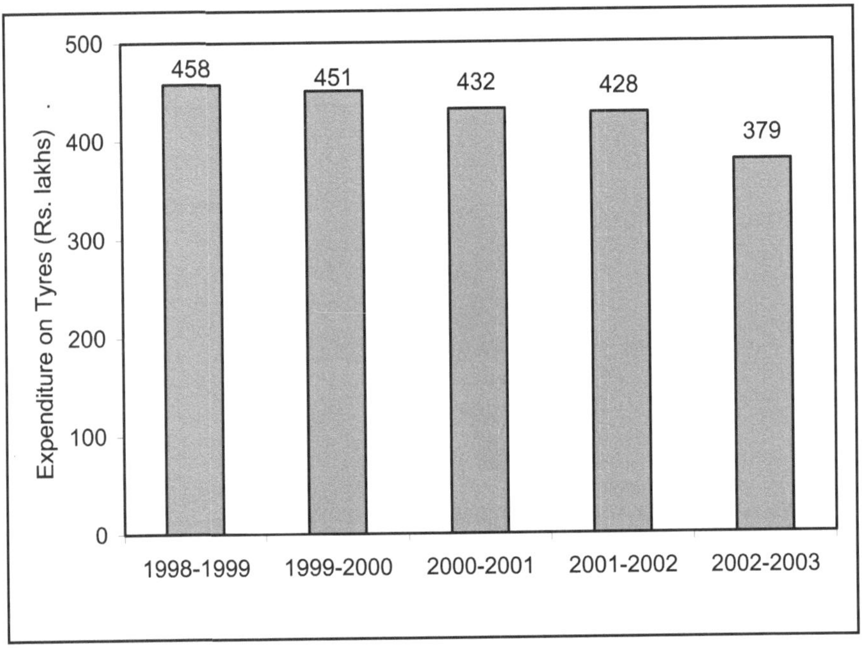

Fig. 2: Yearwise Expenditure on Tyres

3. Brand of Tyres Used

The tyres used at this mining project are procured from different manufacturers like:

- Goodyear
- MRF
- Dunlop
- Firestone
- Yokohama
- Fig. 3 shows the different types of tyres used at this mining project.

Fig. 3: Types of Tyres

4. Maintenance of Tyres

In order to take care of the costly tyres, the maintenance group adheres to the following:

- Daily checking of air pressure
- Weekly condition monitoring of tyres to ascertain the wear pattern
- Painting of rims to prevent rust and help in avoiding air leakage
- Use of standard accessories like valve core, valve caps, valve extension, o-rings, inflation gauges
- Use of tyre dismantler.

The project has developed the following facilities for tyre maintenance:

- Tyre handlers for easy removal and fitment of tyres
- Tyre dismantler for dismantling and assembling of tyres with rim.

Fig. 5 shows the improvement in dismantling and assembling practice of tyres at the maintenance section of the mines as compared to the earlier approach (Fig. 4).

Fig. 4: Old Practice

- Unsafe
- More Manpower involved
- Damage to tyre shoulder

Fig. 5: New Practice

- Safe
- Less Manpower involved
- No damage to tyre shoulder

5. Enhancing Tyre Life

In order to enhance the life of tyres, certain measures are to be taken in practice as listed below:

- Procurement of tyres of correct specification during maintenance replacement.
- Weekly inspection of tyres to understand the requirement of retreading.
- Use of rock ejector and rock guards on the earthmoving equipment
- Maintenance of haul roads
- Grading of roads daily in the monsoon season.

A statistical study on the life of new tyres used in mining equipment is presented in Fig. 6.

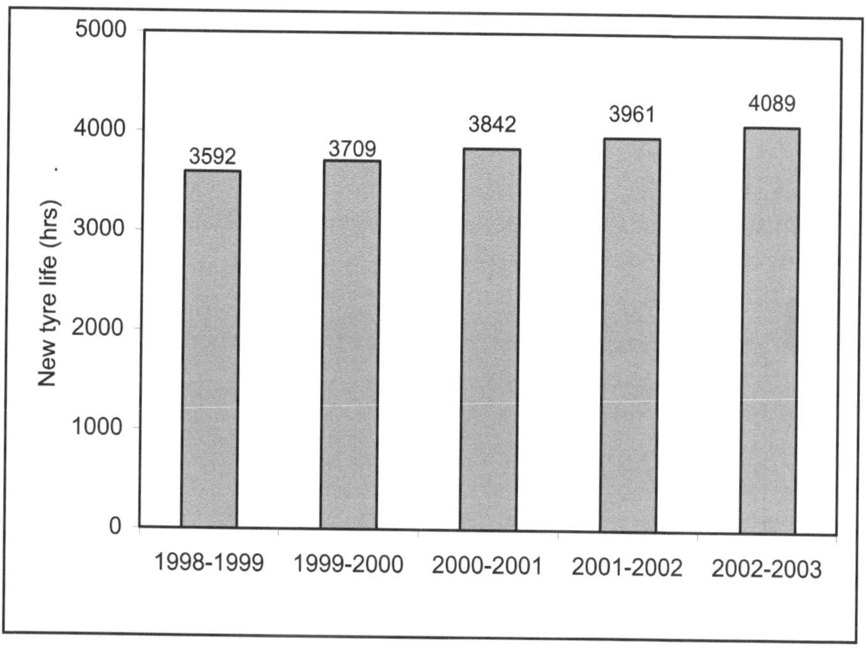

Fig. 6: Life of New Tyres

6. Retreading of Tyres

- The basic objective for the retreading of tyres is to enhance the life of tyres.
- In order to understand the need for retreading, it is necessary to go for weekly measurement of the tread depth of tyre.

- An internally developed software is available using which it would be possible to decide the need for tyre retreadng. The basic features of this software are shown in Fig. 7.
- The removal of tyres from the earthmoving equipment is always recommended before there is any occurrence of tyre failure.
- Further, it is necessary to monitor the life of retreaded tyres also.

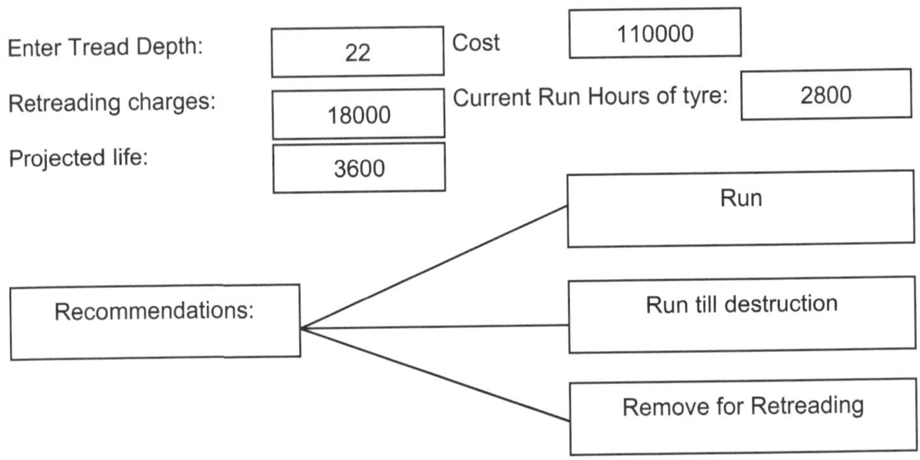

Fig. 7: Features of Tyre Retreading Software

Fig. 8 shows the yearwise trend in increase of percentage of tyres retreated at the West Bokaro opencast mining project. The trend in variation in the life of retreated tyres is shown in Fig. 9.

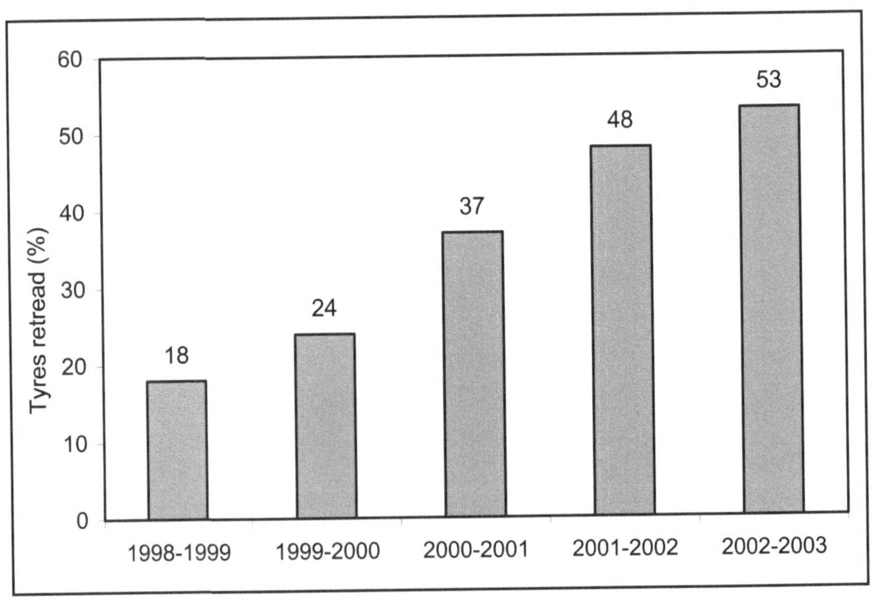

Fig. 8: Statistics of Tyres Retreaded

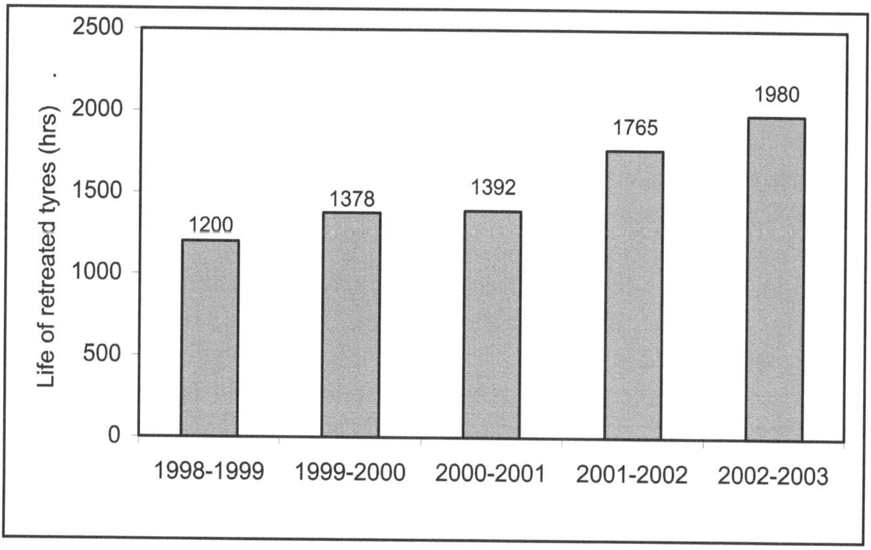

Fig. 9: Life of Retreaded Tyres

7. Tyre Failure

The reasons for tyre failure are listed below:

- Through cut
- Improper inflation
- Uneven loading
- Ply separation
- Spinning of tyre
- Long gradient negotiation

Fig. 10 represents the decreasing trend of failure of the tyres due to through cut.

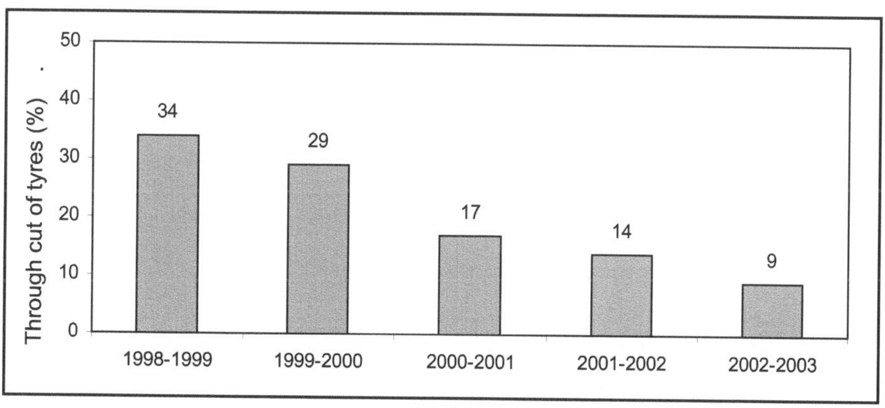

Fig. 10: Tyres Failed due to Through Cut

8. Measures to Reduce Tyre Failures

The reduction in maintenance replacement cost would be possible when the failures of tyres are reduced. In this connection, necessary measures should be taken as are listed below:

- Improved condition of haul roads, dump yard and face.
- Proper drainage system.
- Provision for banking to avoid drastic shifting in load.
- Smooth turns would help to reduce shifting of load on any particular tyre.
- Proper blasting and fragmentation of rocks to avoid any large concentrated load buildup.
- Provision for display boards for speed limit, gear range and use of retarder.

The project has a tyre management system that functions in the manner given below:

- Tracking of tyres has to be done by tyre number so that the tyres can be tracked throughout its life.
- Implementation of Bin-card system for each tyre.
- It is necessary to have analysis of tyre life on the basis of manufacturer, ply rating and grade negotiated.

The project keeps a close eye on the scrap tyre pile to take necessary measures for tyre disposal as per steps given below:

- To have separate tyre yard for scrap tyres.
- The tyres with sidewall cut or shoulder damage or ply cut to be sent to tyre yard.
- Cross checking of failed tyre to be done by retreaders.
- Reverse auction system followed for tyre disposal.
- Only certified parties are authorized for participation in such auction process.
- Tyre numbers are recorded for the tyres auctioned.

9. Challenges Ahead

The mining project envisages to implement the following facilities for the maintenance and management of tyres:

- Online monitoring of tyre pressure
- Reduction in tyre failure due to through cut
- Technology upgradation for sidewall and shoulder cut repair
- Retreading of radial tyres
- Prevention of air leakage through pin holes.

10. Scope for Improvement

The scope for improvement in tyre management system should be examined in the light of following:

- Switching over from cross ply type to radial tyres
- 2nd and 3rd cycle of retreading of tyres
- Annual Maintenance Contract for tyre management
- Retreading of tyres by the concerned tyre manufacturer
- Reduction in tyre expenditure.

Dumper Tyre Failures on Openpit Haulage Routes: A Case Study

S.S. Gupte and I.L. Muthreja

Department of Mining Engineering,
Visvesvaraya National Institute of Technology, Nagpur-440011

1. Introduction

Tyres are the major elements contributing to the operating cost for off-highway mining trucks or dumpers. Expenditure for operating the dumper fleet in opencast mines may be in the range of 30-50% of the total cost of mineral production. It has been observed through various studies that the major components of haulage cost are fuel costs (18–22%), salary of operators (20–30%) and tyre costs (18–22%). Also the tyre cost may vary depending on the model of dumper. Hence, it is essential to maximize the operating life of each tyre. Factors that affect the service life of tyres are:

- Working shift (day or night)
- Practices followed by truck and shovel operators
- Truck make and type
- Speed of loaded and unloaded truck
- Haul route design and construction
- Intensity of using the individual haul routes
- Haul route maintenance practices
- Horizontal haulage distance
- Up the ramp haulage distance
- Ramp gradient
- Tyre construction
- Tyre rotation practices.

The effects of the above factors are well understood. The variable, which is closely related to tyre life, particularly over longer haul length, is the truck productivity as measured in ton-kilometer per hour. The influence of individual haul routes on tyre failures would be difficult to quantify in practice. The failures mainly associated with a particular haulage route may come from:

- High traffic density
- Poor haulage route maintenance
- Poor shovel cleanup practices
- Poor design of haulage routes.

This paper would discuss on the scope for application of potential statistical techniques to analyze the data of tyre failure to be able to correlate the tyre failure with haulage routes in opencast mine. The methodology can be used to identify those haulage routes that have a high incidence of tyre failures. Based on such study, it would be possible to modify or improve the haul road maintenance practices. This study has considered only two types of tyre failures i.e. impact failure of tyres and ply separation.

2. Methodology

Data collection has been done from a particular mine in two different sets, while each data set is for a duration of five months. Each such set consists of data for ply separation failure and impact failure of tyres.

The last row of each of the above tables summarizes the number of times a particular haul road gets associated with a failure using the given data set. It can be seen from Table 1 that haul road 3 is associated with 4 of the 9 ply separation failures, whereas it may be observed from Table 2 that the haul road 5 is associated with 5 of the 10 ply separation failures.

Table 1

Failure No.	*Haulage Road Associated with Ply Separation Failure for the Period January–May 2000*							
	1	*2*	*3*	*4*	*5*	*6*	*7*	*8*
1.	-	-	1	-	-	-	-	-
2.	-	1	1	-	-	-	-	1
3.	-	-	1	-	1	-	-	-
4.	1	-	1	-	1	-	-	-
5.	-	-	-	-	-	-	-	-
Total	1	1	4	0	2	0	0	1

Table 2

Failure No.	*Haulage Road Associated with Ply Separation Failures for the Period September–January 2001*							
	1	*2*	*3*	*4*	*5*	*6*	*7*	*8*
1.	-	-	1	-	1	-	-	1
2.	-	-	1	-	1	-	-	-
3.	-	-	-	-	1	-	-	-
4.	-	-	-	1	1	-	-	-
5.	-	-	-	-	1	-	-	-
6.	1			-	-			-
Total	1	0	2	1	5	0	0	1

If the failure counts are assumed to satisfy Poisson distribution (generally used to model random arrivals during fixed time intervals), it would be possible to test for significant differences between the observed count rate and desired count rate or target count rate using the F-test on the basis of following approach.

Let μ_1 and μ_2 denote the means of two populations of failure rates, where N_1 and N_2 failures are observed over different time periods, t_1 and t_2, respectively. To test the hypothesis that $\mu_1 = \mu_2$,

$$F = [1/ t_1(N_1 + 0.5)]/ [1/ t_2(N_2 + 0.5)]$$

can be calculated, which follows the F-distribution with $(2 N_1 + 1, 2 N_2 + 1)$ degrees of freedom. This statistics is now compared with the critical value of the F statistics obtained from the standard statistical tables available. The null hypothesis is rejected if the critical value is exceeded.

Considering the data in Table 1, the maximum rate observed ($N_1 = 4$ associated over $t_1 = 5$ months) can be compared with the average expected rate. In total, 9 separation failures were observed during a period of five months. Therefore, $\mu_1 = 9/5 = 1.8$.

The maximum observed rate can be compared with the average expected rate considering the data from Table 2. In total, 10 separation failures were observed during a period of five months.

Average no. of haul roads used before separation failure = 3

Now, $N_2 = (3 \times 9)/8 = 3.375$ approximately 3.

$N_2 = 3$ over $t_2 = 10$ months period.

$V_1 = 2 N_1 + 1 = 2 \times 4 + 1 = 9$ (Degrees of freedom)

$V_2 = 2 N_2 + 1 = 2 \times 3 + 1 = 7$ (Degrees of freedom)

Now,

$$F = [1/ t_1(N_1 + 0.5)]/ [1/ t_2(N_2 + 0.5)]$$

$F = [1/5(4 + 0.5)] / [1/10(3 + 0.5)] = 2.57$ with $V_1 = 9$ & $V_2 = 7$ degrees of freedom.

The critical value of F evaluated at 90% confidence level is 2.51. Hence, sufficient evidence exists to correlate haul roads with tyre separation failures.

3. Analysis of Impact Failures

The same methodology can be applied to analyze the influence of haul roads on tyres that fail through impact. As the impact failures occur suddenly, the haul roads used immediately before failure would be of primary interest. Fig. 1 & 2 show the haul roads associated with impact failure of tyre for the period of five months. Impact failures frequently occur in loading and dumping areas.

Total number of impact failures during January–May, 2000 is 10, $N_1 = 6$ over a period of 5 months, $t_1 = 5$ months.

Total number of impact failures over a period of 10 months (January–May 2000 & September 2000 to January 2001) are 24 over a period of 10 months.

$N_2 = 24/6 = 4$ and $t_2 = 10$ months.

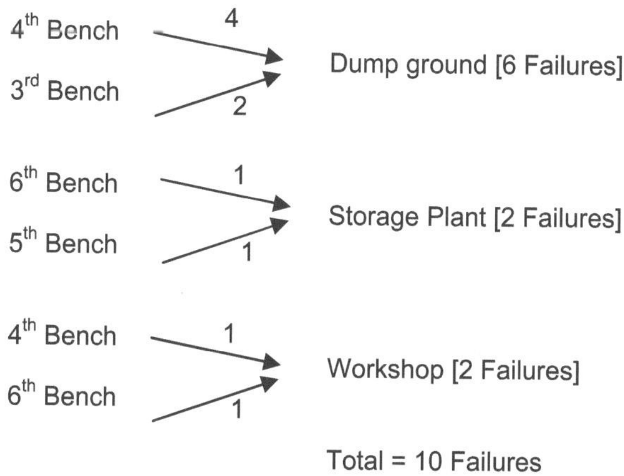

Fig. 1: Impact Failure from January to May, 2000

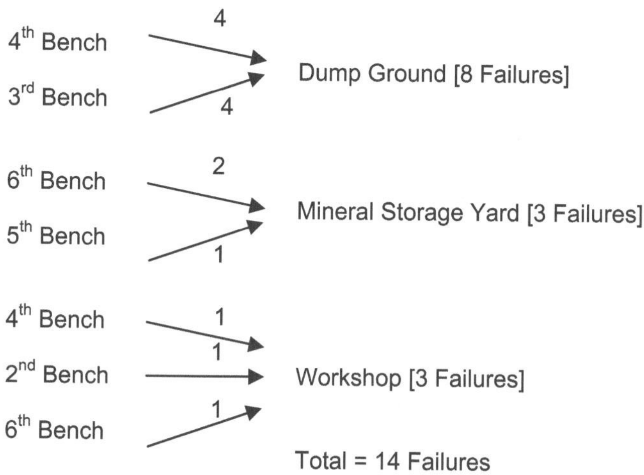

Fig. 2: Impact Failure from September 2000 to January 2001

Maximum number of impact failures during January to May, 2000 i.e. $N_1 = 6$ over a period of 5 months, $t_1 = 5$ months. Total number of impact failures observed during 10 months (January to May, 2000 & September, 2000 to January, 2001) are 24;

$N_2 = 24/6 = 4$ and $t_2 = 10$ months.

$V_1 = 2\ N_1 + 1 = 2 \times 6 + 1 = 13$ (Degrees of freedom)

$V_2 = 2\,N_2 + 1 = 2 \times 4 + 1 = 9$ (Degrees of freedom)

$F = [1/\,t_1(N_1 + 0.5)]\,/\,[1/\,t_2(N_2 + 0.5)]$

$F = [1/5(6 + 0.5)]\,/1/10(4 + 0.5)] = 2.42$

Value of F = 2.42 with V_1 = 13 & V_2 = 9 Degrees of freedom. Value of F at 90% confidence level (from standard table) = 2.16. Therefore, it can be concluded that there is sufficient evidence of tyre failures due to impact.

4. Recommendations

In order to increase the tyre life and reduce the incidences of separation failures as well as impact failures on haul roads, dump ground and mineral storage yard, necessary actions would be required including some changes in cleanup practices at the dump ground and mineral storage yard. The recommendations in this context are:

- Tyre rotation practices may be changed.
- The turning speed may be reduced.
- The turning radius may be increased.
- To identify the haul roads most sensitive to failures.

5. Conclusions

Dumper tyre cost represents a significant portion of the total mining cost in opencast projects. Hence, the scope for increasing the tyre life should be given due priority. In this context, it is necessary to take up studies on site specific factors influencing the tyre life. Necessary data should be gathered and analysed to find out the causes of tyre failures. Assuming that the failure counts satisfy a Poisson distribution, F-tests can be applied to determine significant deviations from the desired level.

6. References

[1] Bridgestone Corp., "Maintenance: Off-The-Road Tyres," publication no. 0ORA02E, Tokyo, Japan, 1990.

[2] Chandwick, J., "Haul roads," *Mining* Magazine, July, pp. 30–33, 1996.

[3] De, A., and Mukhopadhyay, A.K., "Selection, maintenance and relations of various parameters for off-highway hauling tires," *Off-Highway Haulage in Surface Mines*, Golosinki and Srajer, eds., A.A. Balkema, Rotterdam, pp. 153–159, 1989.

[4] Knights, P.F., and Boerner, A.L., "Statistical correlation of off-highway tyre failures with openpit haulage routes," Mining Engineering, August, pp. 51–56., 2001.

[5] Kangi, G.K, 100 Statistical Tests, SAGE Publications, London, 1999.

How to Get Better Tyre Life for Economic Earthmoving

U. Das and W.P. Mahapatra

Jayant Project, Northern Coalfields Ltd.
Sidhi-486890

1. Introduction

Earthmoving and mining tyres have become increasingly important commensurate with the developments of large construction vehicles. The OTR tyres are mostly used in vehicles that are used for load-and-carry operation, i.e. the primary task of these heavy-duty tyres is to haul heavy loads faster and over longer distances. The main constraints to tyre life are generation of heat and premature wear and tear. In order to get the best services from the tyres when used for load-and-carry operation, the following should be taken into consideration:

2. Operation

Operation of the heavy vehicles is closely related to good maintenance practices to help in avoiding the causes of premature failure of tyres. The following are some operational factors that are to be monitored for prolonged tyre life.

- To avoid sharp turns.
- To avoid overloading.
- To avoid abrupt starts and stops.
- To avoid rough and hard operation.
- Not to exceed the speed limit in driving.
- To limit to maximum permissible inflation pressure.
- To avoid stones and rocks.
- To avoid oily and greasy road surfaces.
- To avoid sudden and hard braking.
- To minimise wheel spinning.
- To avoid overloading and spillage of load.
- To limit to permissible inflation.

3. Role of Ton-Kilometer-Per-Hour (TKPH)

A heavy duty haulage causes heat development in tyres. As the tyres have limited resistance to heat, deterioration of tyres may begin at an early stage of operation due to excess heat development. Hence, it is necessary to determine the amount of work that will keep tyres within a safe range to avoid overheating when the vehicle is operated under a given set of conditions. The amount of work to be done under the given conditions and within the safe range can be determined as below:

TKPH = Mean-tyre load × Average work-day speed

where,

Mean tyre load = (tyre load empty + tyre load loaded)/2

Average work-day speed = (Round trip distance × number of cycles/day)/total hours of operation/day

The "Ton-Kilometer Per Hour" (TKPH) is of great assistance in determining the maximum life of a tyre in operation. The following factors should be borne in mind with regard to TKPH

(a) Maximum operating TKPH should not exceed the tyre TKPH rating. If the calculated TKPH requirement is higher than the TKPH rating of the tyre being used, it would be necessary to reduce speed and/or load, else a tyre of higher TKPH rating should be used.

(b) The average operating TKPH should be calculated after several checks and the same should not exceed the tyre TKPH rating.

(c) The TKPH rating data is applicable in relation to the following:
   - Suitable for all standard, heat resistant and cut resistant tyres.
   - Ambient temperature to be less than 38°C.
   - Maximum trip distance to be less than 5 Km but extendable to 16 km with reduced load.
   - Maximum speed to be 50 km/h.

The TKPH is very much influenced by heat. When the heat in tyre increases due to abnormal conditions like high ambient temperature, speed and overloading, the TKPH should be reduced. Similarly, adjustments to TKPH can be made if the ambient temperature or the speed is low. The load carried by the individual tyre must be in line (Fig. 1).

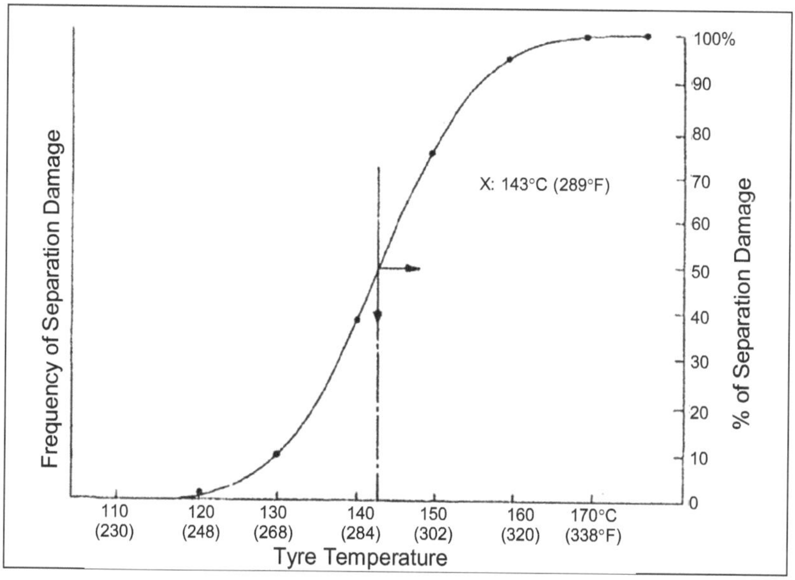

Fig. 1

The average tyre load and the average operational speed for obtaining the maximum amount of work using a given type of tyre can be obtained from the operational capacity graph of tyre as shown in Fig. 2.

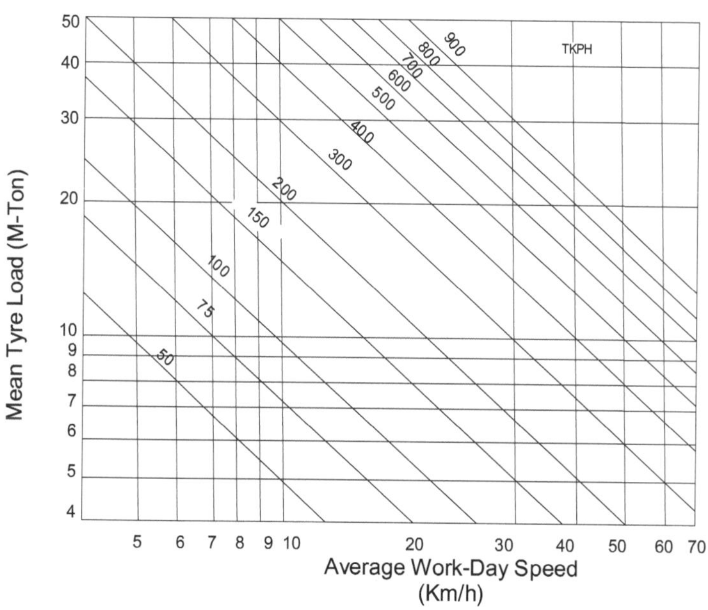

Fig. 2

3. Loading

The load carried by transporting equipment is also a major factor in producing heat in tyres. So each tyre fitted on the vehicle should be within the recommendation limit for the particular tyre size, ply rating and inflation pressure. The load is to be determined after the calculation of TKPH and should not exceed beyond the predetermined value.

3.1 *Damage Due to Overloading*

Overloading a tyre result in the combined effect of over and underinflation that the tyre has to experience. The tyre is susceptible to excessive tread and sidewall movement, cord tension and bead stress resulting in excessive heat generation. Consequently, the tyre life will be adversely affected and early life failures like separations, impact fractures, penetrations and reinforce flex breaks may be expected. It has to be remembered that:

- Overloading reduces the life of a tyre.
- Overloading will cause permanent tyre damage.

3.2 *Problems Caused by Overloading*

Overloading may cause various problems as listed below and as a precautionary measure, the vehicle speed should be reduced subsequently.

- Ply separation due to excessive distortion and increased stress on sidewall.
- Cord and bead damage due to excessive deflection.
- Uneven wear due to excessive tread movement.
- Cut and impact damage due to increased cord tension.
- Damage of vehicle due to excessive heat generation.

The tyre performance due to excessive loading is shown in Fig. 3.

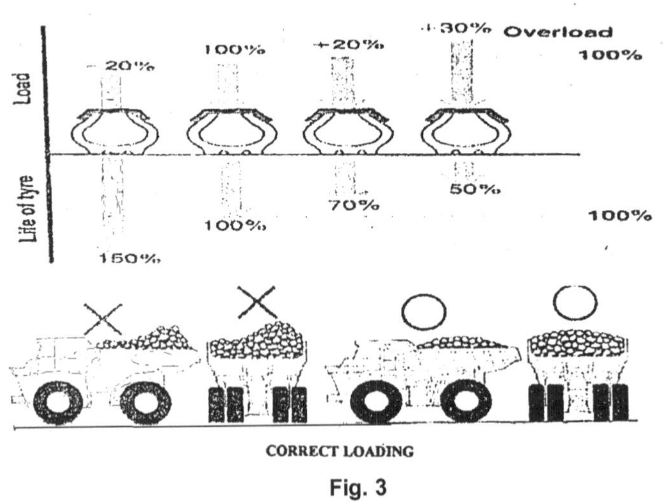

Fig. 3

4. Effect of Speed

The speed of the vehicle should be decided on the basis of prevailing conditions e.g. climate, haul road conditions, gradient and visibility. The following aspects should also be kept in view in this connection:

4.1 *Damage Due to Excessive Speed*

An excessive speed of the vehicle would result in higher heat generation due to increase in the rate of flexing that takes place. This can result in tyres failing due to heat separations, impact fractures and penetrations. The life of the tyre will be adversely affected due to increased rate of wear caused by excessive braking and sliding while cornering the vehicle.

4.2 *Correct Working Speed*

The correct speed is determined from both the vehicle's maximum speed limit range and the conditions prevailing at the site of work.

The recommended maximum speeds under ideal conditions are shown in Table 1.

Table 1

Type of Vehicle	*Distance*	*Maximum Speed*
Earthmovers	5 km	50 km/hr
Graders	Unlimited	40 km/hr
Dozers/Payloaders	600 m	10 km/hr

The average life of tyre in percent with reference to loaded speed of the vehicle and grade of roadway is presented in Fig. 4.

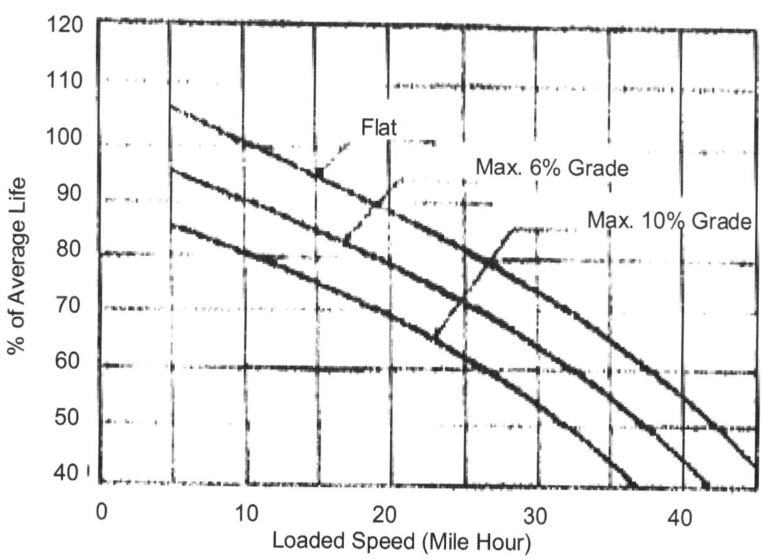

Fig. 4

5. Road Maintenance

Off-the-Road Tyres are used on unpaved roads and these are often littered with objects which could easily damage a tyre. Hence, the haul road meant for the vehicles fitted with OTR tyres should be maintained properly.

5.1 *Steps for Proper Haul Road Maintenance*

- Constant care should be taken for the haul roads and any obstruction on the road should be removed immediately.
- The road surface must be graded uniformly to improve drainage. Further, filling up of holes, elimination of corrugations and removal of the partially protruding rocks are the essential requirements.
- The haul roads must be watered regularly, but not excessively, to reduce the dust hazards.
- Super elevation (or banking up) of haul road corners is recommended to eliminate load transfer during cornering.
- Steeper gradients would cause tyre slippage and would result in rapid tyre wear, increased fuel consumption and reduced work efficiency. The gradient of the haulage road should never exceed 10%.
- Correct drainage of haul roads is essential to prevent the accumulation of water on the road. The road contour should have 3% crowning and ditches should be made at the sides to allow for drainage and water dispersion (Fig. 5). The haul road should be built to be 4 times the width of the vehicles plying on the road to and fro.

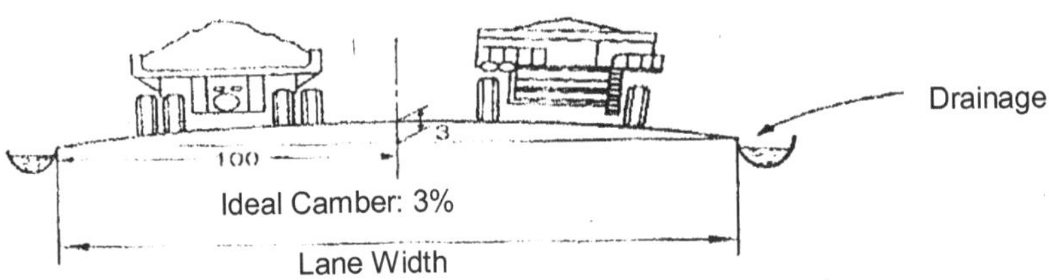

Fig. 5: Road camber and lane width

6. Vehicle Maintenance

Proper maintenance of the vehicle is absolutely necessary for its safe and economic operation. The use of vehicles not maintained properly is the major cause of accidents and therefore such vehicles should not be used. The following should be checked to help in prolonging the life expectancy of tyres:

- A careful checking of the axles is essential. A bent or sprung axle can result in uneven load distribution and is likely to lead to alignment problems. This would also result in an irregular wear.

- Installation of rock guards on the dumpers would help to prevent tyre damage due to spillage. The stone ejectors between the dual tyres are essential as this would help to prevent the stones from becoming wedged between the tyres.
- Necessary checking should be done for oil, grease and fuel leakages. The tyres absorb petroleum products and this softens the rubber as well as result in swelling and aging of rubber. As a result of this the tyre life is shortened.
- The brakes of the vehicle must be checked regularly. An unequalised braking would cause a wheel to get locked prematurely resulting in excessive tyre wear.
- The wheel alignment should be checked regularly, especially for the front wheels. Any misalignment would also cause uneven tyre wear.

7. Tyre Maintenance

Generally, the tyres in surface mining operation go as the next high cost component after the fuel cost. In view of this, a strict tyre maintenance programme should be instituted; otherwise it would increase the cost parameters and the equipment downtime both. The maintenance of tyres should involve a schedule of daily inspection.

7.1 *Tyre Inspection*

Off-the-road tyres are subjected to abnormal abuse due to the severity in operating conditions. For safety and economic reasons, the daily inspection schedule should take care of the following.

- Inflation pressure
- Tyre appearance
- Tyre valve and rim components and their appearance
- Tyre storage
- Tyre transporting
- Proper matching of tyres.

7.2 *Tyre Inflation Pressure*

While the tyre is in operation, considerable heat gets generated. This causes the pressure in the tyre to increase. Sometimes, the pressure increase is in excess of 25% of the cold inflation pressure. The tyre must not be bled to reduce the pressure increase. If the pressure increase is in excess of 35%, the tyre should be allowed to cool and be rechecked, or it be removed for inspection.

The inflation pressure should be checked daily before using the vehicle. If this is not possible, 'hot pressure' of the tyre should be recorded daily. An analysis of the pressure recordings will show possible leaks and other problems in tyres before these are getting damaged badly.

The inflation pressure must be checked and corrected whenever necessary before operating the vehicle. This ensures that the tyres start working at the recommended *Cold Inflation Pressure* as stipulated by the manufacturer. The trend in variation of inflation pressure is presented in Fig. 6.

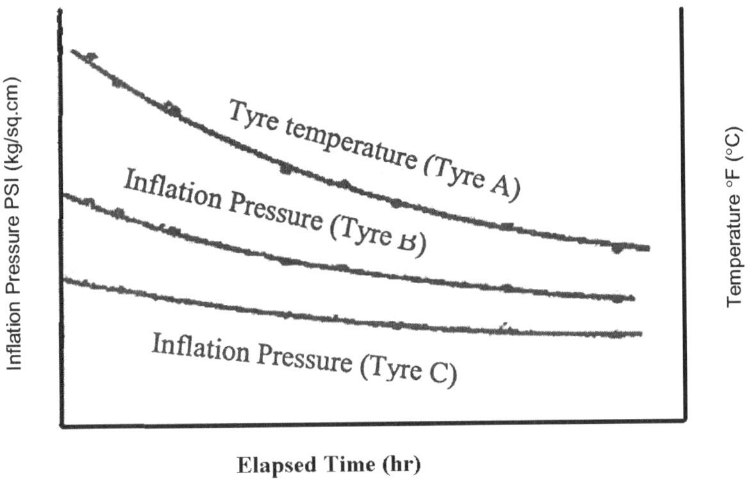

Fig. 6

7.3 *Tyre Appearance*

A visual inspection of all the tyres should be carried out daily. An early finding of any tyre damage can reduce expensive tyre repair at a later stage. Some of the ways of reducing the losses from a damaged tyre are listed below in Table 2, in case the same is detected at an early stage.

Table 2

Type of Problem	*Action Needed*
Foreign matter in tread	Remove and repair
Separation	Remove and inspect
Tread cuts into the body	Remove and repair
Tread groove cracks	Check inflation history and run vehicle
Uneven wear	Check alignment and inflation
Oil impregnation	Clean
Tyre touching the vehicle	Report to maintenance section and get the condition inspected

Daily inspection would help to highlight the mechanical defects so that necessary corrective actions can be taken at the earliest.

7.4 *Tyre Valve and Rim Components*

The tyre rim components play a vital role in relation to performance of a tyre. Leakages due to faulty rim components will cause excessive heat build-up due to under inflation. In case the same is allowed to remain so, the tyre will fail due to the 'run flat' condition. Daily checks of the valve caps and extensions as well as the rim must be done during the time of pressure checking.

7.4.1 *Rim Maintenance*

- The rims must be free from any rust. If rusted, the rust should be removed by wire-brushing or sand blasting. After it is free from rust, painting of the same should be done with an etching primer.
- The rims should be handled carefully to avoid damaging of bead seats and distortion of rim-flanges.
- Care should be taken not to allow any damage of the valve hole.

7.4.2 *'O' Rings*

- O-rings should be stored in a cool and dry place.
- A new O-ring should be used whenever a tyre is mounted on the rim.
- It should be ensured that the O-ring of correct size is used.

7.5 *Tyre Storage and Transportation*

7.5.1 *Storage*

Tyres in stock should be grouped as new, used and worn out tyres as per the sizes and the types of tyres. Tyres held for matching should be classified by crown circumference or remaining skid and should be grouped properly for future use. Wherever possible, the tyres should be stored in a cool and dry place. Outdoor storage of tyres is not recommended. Tyres should be stored upright and in case the same is not possible, the tyres should not be stacked in height more than three times the thickness of each tyre. Storing of tyres near electrical installations and near the petroleum products should be prevented. Wherever possible, the tyres should not be stored on mothballed vehicles. If this is unavoidable, the vehicle should be stored on blocks with the tyres in suspended position. The tyres must be coated at regular intervals of time with Bridgestone or Firestone protective paint and the same should be protected from any exposure to open weather.

The tyres should be grouped as below:

- New tyres (not mounted on rims)
- New tyres (mounted on rims)
- Matching tyres (grouping by circumference or by remaining skid)
- Demounted tyres (unexamined)
- Demounted tyres (examined)
- Scrap tyres.

7.5.2 *Transportation*

When transporting the tyres, care should be taken to ensure that the beads and inner liner of the tyre are not damaged. Any minor damage of tyre may even result in premature tyre failure due to air seeping into the body of the tyre causing separation. The correct and incorrect methods of transporting a tyre are illustrated in Fig. 7.

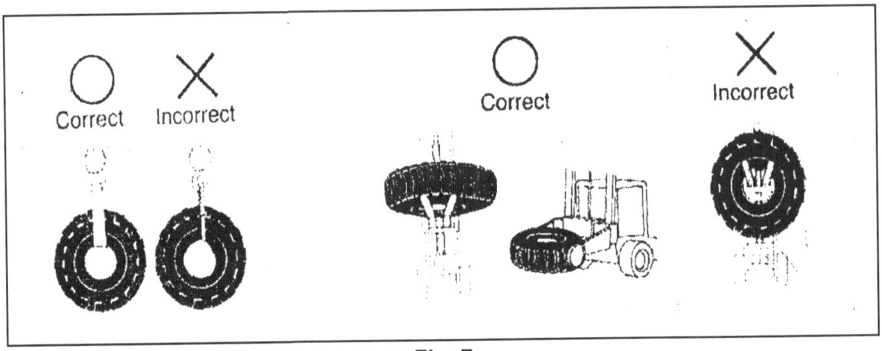

Fig. 7

7.6 *Matching of Tyres*

As a general rule, when two tyres are to be fitted on the same axle, the tyres of the same type and of same overall diameter should be chosen as a preference. Mismatched combination will result in rapid tread wear on the smaller tyre and mass displacement on the larger tyre. A careful attention with regard to the following would assure the maximum serviceability of tyres.

7.6.1 *Tyre Matching Procedure*

- Dual mounting of the same brand of tyre is ideal; if this is not possible then tyres of different brands, but of the same type and construction and of the same diameter should be used together.
- Dual mounting of tyres with different section widths is not recommended, even if, they meet minimum values of dual spacing, because of differing deflection.
- Cross ply and radial ply tyres must not be mounted on the same axle.

Dual tyres if mismatched, the tyre with larger outside diameter will eventually show signs of rapid and uneven wear due to overloading, while the tyre with the smaller diameter will wear out in the centre because of scuffing.

There is a wide variety of the OTR tyres that are used in India. Although in most of the operations the OTR tyre is the next cost component after the fuel cost, the productivity and the cost per unit would depend on the tyre performance. A proper tyre management is yet to be taken up in this country and the sooner it takes place, the better will be the future.

The Use of Off-Highway Tyres, Its Importance and Earthmover Management System (EMS)

S.S. Gupte and I.L. Muthreja

Department of Mining Engineering,
Visvesvaraya National Institute of Technology, Nagpur-440011

1. Introduction

In opencast mines, the mine operators are using a good number of off-highway tyres and spend heavily on them (Table 1). There are some basic considerations to extend the tyre life and lower the costs. It becomes mandatory to recognize the technical limitations of tyres and proper operating procedure should be followed that would avoid mishandling and improper use of tyres. All these could be financial and performance drawbacks for many operating mines.

Table 1

Truck Population as per Mineral Market			
Mineral	*No. of Mines*	*No. of Trucks*	*% of Total*
Coal	174	2537	38.7%
Copper	52	1676	25.6%
Iron	48	791	12.1%
Gold	42	762	11.6%
Others	48	786	12.0%
TOTAL	364	6552	

The overall tyre expenses can be upto 20% of haulage costs, whereas the haulage cost includes both owning and operating costs. The operating cost is about 75% and the tyre cost can be upto 25% of that depending on the mining conditions. When presented in other form, it may go for spending so much on tyres over the life of a dumper, as much it was for purchasing the same haulage truck at the first instance.

Tyre management and tyre selection is important right from the start of operation. Tyres are becoming more application specific with regard to considering tyre tread and rubber compound. Also the site conditions, operational practices for the vehicle and maintenance conditions should be duly considered. Most of these factors are controllable and can be managed properly to enhance the tyre life.

Extreme environmental conditions as prevalent in openpit mines can cause a havoc on tyres and can affect the operating cost of the mine largely. In order to avoid this, the mine operators are increasingly interested in some kind of monitoring systems that can alert them when the tyres are operating outside the range of design parameters.

2. Selection of Tyres

The important functions of a tyre can be summarized as below:

- Supporting load,
- Transporting, traction and braking on the road surface,
- Absorbing shock from road,
- Changing driving direction and sustaining it.

Proper tyre selection is the basic requirement to keep the operating cost at a minimum. The major factors involved in the selection of tyres are speed, road surface and continuity of operation. In order to minimize the mining costs, the tyre maintenance operators should aim as below.

- Reduce operating cost by prolonging tyre life,
- Reduce equipment downtime by minimizing tyre damage,
- Increased safety.

2.1 *Tyre Inflation Pressure*

Proper tyre inflation is one of the most important factors to enable prolonged tyre life. The proper inflation pressure depends on vehicle type, size, ply rating, actual vehicle speed and tyre load. Incorrect inflation pressure goes for lost tyre value. In one case, about 80% of tyres in mines were within +/– 10% of recommendation and the site was losing 8% of tyre value by not having accurate enough air pressure management. The recommendation was to institute a system of daily pressure checks.

Overinflation results in higher ground bearing pressure due to reduced contact area. This results in excessive wear at the crown area of the tyre, reduced traction and flotation.

Underinflation causes excessive tyre deflection and distortion in the sidewall area that leads to uneven tread wear, radial cracks, cord fatigue and excessive heat generation. The excessive heat generated will cause the tyre to fail prematurely accompanied by heat separation. Underinflation is more severe on tyres than overinflation.

2.2 *Heat and Tyre Failure*

Tyre manufacturing requires heat in the vulcanizing process to convert crude rubber and additives into a homogeneous compound. The heat required is above 132°C. Also heat is generated as the tyre rolls and it gets flexed. The heat generated is rather faster than that can be radiated to the atmosphere. Thus, the heat gradually builds up within the tyre and reaches the maximum level at the outermost ply. Over a considerable period of time, the heat can develop from overflexing to actually reverse the vulcanizing process and thus ply separation and tyre failure takes place.

2.3 *Overloading and Tyre Life*

Overloading is an important factor that reduces the tyre life. An overloaded tyre experiences the combined effect of over and underinflation.

2.4 *Speed and Tread Wear*

The operator of the vehicle must adhere to the speed limits laid down by the management. The maximum speed recommended should not be exceeded. Excessive speed results in higher heat generation due to increase in the rate of flexing that takes place. It may result in tyre failure due to heat separation, impact fracture and penetration.

2.5 *Ton-Kilometre-Per-Hour Rating*

It is a rating used to determine the maximum amount of work a tyre can safely and effectively handle under given operating conditions without overheating.

2.6 *Rock Cuts*

Apart from the above, the rock cuts have a major impact on cost parameters. There are two areas that should be considered. The number of tyres that go out of service early due to rock cuts and do not deliver the potential life they could have, leading to a loss of potential hours (i.e. how early in their life they were cut). At one site, there is 30% loss of tyres due to rock cuts. Those tyres lost 70% of their life and the site had been loosing about 9% of the tyre value. At another site there is 65% loss of the tyres due to rock cuts. These tyres are lost at 60% of their potential wear life and are losing 20% of their value in the tyre programme. Basically, the tyre costs are about 26% higher than expected. In short, a focus on decreasing the rock cuts will probably lead to an increase in tyre life. At one site the mine was able to decrease the rock cut rate by 6% and thus increase the tyre life by 15%. The total net savings for one year due to the programme was about 7% of that spent on tyres.

3. Maintenance Tips

Since tyre prices are very high, it is the time now to follow some important maintenance tips and to ensure that the tires operate at peak levels throughout its desired life. The key to this is to check the tyre regularly. Routine maintenance reduces downtime, eliminates preventable major repairs, improves operating efficiency and promotes higher level of productivity. Ten simple steps that can save considerable time and money spent in this regard are given below:

Step 1: It is necessary to conduct a visual inspection of the vehicle tyres prior to operation. It is required to look for signs of irregular wear on the tread or shoulder of the tyre. It is necessary to examine the tyre for bubbles or bumps, if any, caused by air infiltration or by foreign objects. If either of these symptoms are noticed, it is essential to have the tyre repaired promptly because both can lead to not only tyre failure but may develop some potential danger.

Step 2: In case deep cracks, cuts or other problems are noticed during the inspection, the vehicle should not be operated. A trained service person is required to diagnose the severity of the problem and make the proper repair. An unskilled person should never be allowed to attempt repairs because incorrectly mended tyres can lead to performance problems in the future or even can result in personal injury whenever the tyre fails.

Step 3: The tyres should be checked for correct air pressures. This job should be performed daily for vehicles in constant use because air pressure is critical to a tyre's performance. The air pressure can be checked weekly for vehicles with less demanding schedules.

Step 4: The vehicle owner's manual should be checked to determine precise air pressure requirements. It should provide initial data on the weight of the vehicle and the standard load. Tyre manufacturer's representative can help to pinpoint the exact air pressure recommendations for the tyres of the vehicle based on manufacturer's recommendations and the requirements at the site where the vehicle is used.

Step 5: The vehicle that has flat tyres, damaged or distorted rims or wheels, missing bolts or cracked studs should never be operated. Any of these symptoms could be dangerous.

Step 6: No welding or application of heat should be done to parts of the wheel near the tyre. Heat causes serious damage to the tyres and can cause them to explode. Tyres should be removed always before these types of procedures are followed.

Step 7: The tyres should be stored properly when they are not in use and preferably be placed in a cool and dry place away from direct sunlight to avoid premature aging. Also any exposure to ozone sources such as sun, arc welding unit, mercury vapor light bulbs, ultra-violet rays and inclement weather should be prevented. The tyres are to be stored as standing upright position on the tread and stacking should be avoided as the practice can weaken the tires at the bottom of the stack.

Step 8: Any lifting of tyres through its center with a crane hook should be avoided as this can damage the critical bead area. Instead, the tyre should be lifted by using flat straps placed under the tread. The flat straps are recommended over steel rope slings or chains as these would not cause cuts or abrasion.

Step 9: The inner and outer tyres of a twin fitment should be deflated before removing any rim fixture from the hub of the vehicle.

Step 10: Mixing of tyres on the vehicle should be avoided i.e. pairing a normal tread depth with another of deep tread depth or a bias-ply tyre with a radial ply, etc. Using two different types of tyres could cause damage to the vehicle's internal components, because the tyres would not work together to provide the same traction and handling performance.

4. Earthmover Management System (EMS)

4.1 *EMS and Utility*

EMS is a real time service tool designed to keep the tyres running in the maximum performance zone. The system can achieve this by continuously monitoring the performance of the tyres and providing critical information on internal air pressure and temperature, while the vehicles are in operation. The system is designed to eliminate the need for field pressure checks to extend the life of the tyres and to decrease the downtime. A survey has revealed that the system can increase tyre life upto 8% depending on the site conditions. Also an accurate monitoring while the vehicles are in service would help the mining project to increase the availability of its vehicle fleet. Obviously, the continuity in operation would be translated to more tonnages moved each day. The operators can feel confident about using the vehicles in service because of the certainty that the tyres are engaged in work as they should. This is extremely important, as accidents caused due to the over inflation of tyres would be significant hazards in the mining industry. With the use of a monitoring system, the operators can be alerted about the potential of any accident to occur. The system would be transparent to the vehicle operators, allowing them to give more focus on the tasks at their hands. Should a tyre require service, the EMS will alert the mine management system and whenever so informed, decisions can be made about the fleet. With the EMS in use and assuming that actions are taken whenever the system displays an alert signal related to a safety issue, the incidents resulting from the tyre overinflation could be prevented.

4.2 *Functioning of EMS*

The system works through the tags located inside the air chamber of the tyre. These tags contain state-of-the-art electronic components and sensors to monitor the inflation pressure and internal temperature. Information from the sensors is transmitted to an on-board receiver mounted at the underside of the vehicle. The on-board receiver serves as the backbone of EMS as it receives data from the tags and sends the same to the center for operations control via an existing mine management system. Once the on-board receiver has relayed critical pressure and temperature data to the operation control unit, the trained operators are able to make well informed decisions about tyre maintenance. Necessary alerts can also be transmitted to vehicle operators with detailed instructions for fleet routing or for service needs.

5. Conclusions

Selection, operation and maintenance of dumper tyres continue to be the most important factors in openpit mining economics. It is essential to maintain the correct inflation pressures and as a result of the same, the tyre operating hours can greatly improve. This operation has a large economic significance for any mines.

The tyre maintenance lays an impact over the entire working site to keep the fleets operating at maximum efficiency. The mine operators can take the advantage of tyre investment and boost productivity levels by following proper maintenance guidelines.

Since it becomes difficult to monitor the condition of so many tyres on running dumpers, necessary consideration may be given for the installation of EMS. The benefits delivered by EMS could be reflected in cost savings, improved productivity and safety. It could also be reflected through increased tyre life by alerting the mine operator whenever the tyres are operating outside the range of their design criteria preventing the costly damages.

6. References

[1] Matthews, A. Patricia and Nowatzki, A. Edward, "The effect of pneumatic tyre characteristics on the performance of off-highway trucks", Mine Planning and Equipment Selection, 1988.

[2] Bridgestone Corp., "Maintenance: off-the-road tyres," 1990, publication 0ORA02E, Tokyo, Japan.

[3] Chandwick, J., "Haul roads," *Mining* Magazine, July 1996, pp. 30-33.

[4] De A. and Mukhopadhyay, A.K., "Selection, maintenance and relations of various parameters for off-highway hauling tyres," *Off-Highway Haulage in Surface Mines*, A.A. Balkema, Rotterdam, 1989, pp.153–159.

[5] Knights, P.F. and Boerner, A.L., "Statistical correlation of off-highway tyre failures with openpit haulage routes," MINING ENGINEERING, August, 2001, pp. 51–56.

[6] Kangi, G.K., 100 Statistical Tests, SAGE Publications, 1999, London.

[7] Gilewicz Peter, "What goes where", World Mining Equipment, October, 2001, pp. 36–42.

[8] Michelin, Maintenance tips ensure tyre performance, March 24, 1999.

[9] Potts, Adriana, "Keep 'em running," World Mining Equipment, May, 2001, pp. 36–42.

Status of Abrasion Resistant Tyres for Mining Industry: A Review

Samir K. Pal and Kaushik Pal

Department of Mining Engineering, Indian Institute of Technology, Kharagpur
Kharagpur-721302

1. Introduction

Generally, the design of engine and other mechanical components receives the major importance for all vehicles, while the tyres are often overlooked. Needless to mention that every vehicle has to run on the tyre. Anyway, the tyre plays a very important role in the automobile industry. The most important application of rubber tyres relates to the transport sector. Tyre industries consume more than 60% of the rubber product and the prime factors of consideration in this connection are safety and tread life of tyre. About 30% of the mechanical work available from the fuel is dissipated at the tyres. A reduction of 15% in tyre rolling resistance may improve the fuel consumption by 4.5%. The basic functions of the tyre are:

- Proven load carrying capacity,
- Provide cushioning and enveloping,
- Transmit driving and braking torque,
- Producing tractive force,
- Provide dimensional stability,
- Resist abrasion,
- Provide steering response,
- Have low rolling resistance,
- Provide minimum noise,
- Permit minimum road vibration,
- Be durable and safe.

Generally, the tyres used in light duty automobiles may not create much problem. However, the heavy duty (with or without tube) tyres used in very large vehicles as in mining e.g. heavy duty loaded trucks, dump trucks, pay loaders etc. create much problem due to the unavoidable damages of tyres on account of abrasion and wear.

Now-a-days, both the tyres and the vehicles are of very sophisticated design. Drivers of these vehicles often forget that the tyres are the one that develop the contact with the ground. However, the tyres need to perform a number of functions such as steering, carrying the load, cushioning, rolling, transmitting the tractive effort and put control on the kinetic energy of the vehicles.

2. Ingredients of Rubber Tyres

The tyres are the essential components in heavy earthmoving vehicles for the transportation of minerals, ores, and rocks. This provides a rapid, economic and safe method of moving the bulk materials from one place to another. Over the past few years, some major changes have occurred in the design of tyres so as to improve the serviceability of tyres even under very rugged duty conditions. Mainly rayon with nylon is added with adhesives to be used for more specialized application. For more durability, the reinforcing agents like polyester, glass-fibre etc. are mixed with the rubber compound. Obviously, rubber is the basic material used in tyre production either in natural or in synthetic form. Natural rubber is obtained from the rubber trees, whereas the synthetic rubber is produced from the by-products of oil and natural gas. Various types of synthetic rubber is now available having lower cost, higher robustness, ease of manufacturing and very low porosity.

2.1 *Carbon Black*

Carbon black is the powdery black pigment produced by incomplete combustion of natural gas. Carbon black strengthens the resistance to wear and tear. It is the reinforcing agent of rubber. Now carbon black is often replaced by long link carbon as the reinforcing agent. The structure of this form of carbon black offers longer life because the tyre surface resists abrasion better. The tyre surface stretches and may develop crack under the action of acceleration, braking and cornering causing tyre wear. However, the long link carbon disperses the cracking energy. The longer links develop collision of fewer molecules and generate lesser heat from friction. They also reduce rolling resistance. Silica is also used in the tread compound as a partial substitute for carbon black. It brings improved wet grip and reduces the rolling resistance still further. One quarter of the total weight of a tyre is made from carbon black when the same has to be used in mining projects.

2.2 *Silica*

The rolling resistance in the use of tyres has been improved through the replacement of some quantity of carbon black with silica in the compound of tyre tread. This has enabled the manufacturers to produce mining tyres with improved wet skid properties, better winter performance and lower rolling resistance. With the addition of silica, the tyre engineers have been able to produce the rubber compound that would have high hysteresis at high frequencies and low hysteresis at low frequencies. The use of silica can result in the reduction of rolling resistance by 20% or more. The use of silica can also improve wet skid performance. Some researchers claim to have improved wet skid performance by as much as 15% due to the incorporation of silica. Eventually, this improves the braking distances considerably. Silica also provides substantial benefits to tyres not only in winter, but in al season. Introduction of silica enhances elastic properties and flexibility at lower temperatures allowing better grip and braking during the winter.

2.3 *Steel and Textile Cords*

Steel, cotton, rayon, nylon, fibreglass and polyester cords are specially treated to make a bond with rubber. These are designed to balance the internal inflation pressure and provide necessary resistance to the strains developed when the tyre is in use for the transportation of solids in bulk.

2.4 *Chemical Composition of Tyre Material*

A wide range of chemical compounds e.g. natural rubber, SBR (Styrene Butadiene Rubber), butadiene rubber etc. can be found in the rubber tyres of vehicles. Chemical analysis of tyre material has revealed that metals like Zn, Fe and Ca can be present in different concentrations. A large variety of other chemical substances are also added to tyre rubber. These are vulcanizing agents, accelerators, retardants, pigments, fillers, reinforcing agents, softeners, anti-oxidants, anti-ozonants and desiccants. Some results from experiments carried out to determine the chemical specification of the tyre material are shown below:

Fauser [1] found the general chemical composition of abrasion resistant tyres as Natural Rubber (NR) 40%, Styrene-Butadiene Rubber (SBR) 30%, Butadiene Rubber (BR) 20%, Butyl and halogenated butyl rubber 10%. Similar work carried out by Hildemann *et al.* [2] showed that the bulk (40-60%) of tyre tread is composed of a variety of rubbers, including natural rubber co-polymers, butadiene rubber, styrene butadiene rubber (SBR), nitride rubber, isoprene rubber, neoprene rubber and polysulphide rubber.

2.5 *Chemical Composition of Tyre Wear Particulate Matter*

Nearly half of the total particulate mass in tyre wear sample is composed of organic compounds with further one sixth being the elemental carbon [2]. A survey has revealed that the zinc concentration is lower than that expected. Some studies on the analysis of tyre wear for heavy earthmoving vehicle dust samples using gas chromatography/mass spectrometry techniques, determined the existence of more than 100 solvent-extractable organic compounds [3]. In such survey, it has been noted that the tyre wear particles contain an appreciable amount of styrene, butadiene polymers and also some Polycyclic Aromatic Hydrocarbons (PAH). A summary of information found in literature regarding the different compounds (organic and inorganic) and their concentrations in tyre wear particles are also reported [4]. These compounds are Elemental Carbon (EC), Organic Carbon (OC), Polycyclic Aromatic Hydrocarbons (PAH), Polymers and Elements e.g. S, Zn, Pb, Cd.

The details of the inorganic compounds present in tyre wear particles [2] are furnished in Table 1.

Table 1: Inorganic Compounds Found in Tyre Wear Particles

Species	*Tyre Wear (% of Mass)*	*Species*	*Tyre Wear (% of Mass)*
Al	0.017	As	0
Si	0.180	Se	0.002
P	0	Br	0.002
S	0.110	Rb	0
Cl	0.052	Sr	0.004
K	0.038	Ba	0.037
Ca	0.2	Pb	0.016
Ti	0.056	EC	15.3
V	0	OC	36
Cr	0.003	Mg^{+2}	0.036
Mn	0.010	Na+	0.068
Fe	0.460	Cl	0.060
Ni	0.005	NO^{3}	0.150
Cu	0.049	$SO4^{-2}$	0.250
Zn	0.043	NH_4^{1}	0.019

3. Construction of Rubber Tyres

The main components of the mining tyre carcass are (a) Plies and sidewall rubber, (b) Beads and the components of the tread area are (c) Tread rubber (d) Belt package.

3.1 *Casing, Plies and Sidewalls*

Fabric means the body of tyre that gives its strength to hold the internal pressure for supporting the load. The polyester is the most widely used carcass material. The carcass of most of the radial-ply tyres of commercial vehicles and of some light trucks have steel cord plies. In most cases the load carrying capacity of radial tyres is governed by a load index, which is presented in the form of a table consisting of numbers with the corresponding maximum load per tyre in kilograms.

The sidewall means the sides of the tyre i.e. between the tread and the bead. Technically, one may refer only to the protective rubber outside providing the coating on the sides. Sidewall requires the rubber that is supple enough to allow flexing; yet provide the protection against scuffing and cutting. A blend of Styrene-Butadiene rubber is used for such purpose.

3.2 *Beads*

The tyre beads are the bundles of strong steel wires that prevent any change in shape, which would influence the pressure that supports the load. The beads on a tubeless tyre are made to fine tolerance to maintain the airtightness, where the rim and tyre seal. The bead wires are encased in hard triangular rubber known as "bead fillers or flippers". The flipper is designed to provide a smooth transition from the very stiff bead area to the highly flexible sidewall area. The bead heel is the part of bead that fills the area where the rim flange and the bead are located. Special rubber is used to protect the tyre from chafing against the rim.

3.3 *Tread Rubber*

It is the part of tyre that comes in contact with the road. It provides traction, cushioning quality and cut resistance. The most obvious characteristics required for the tread rubber is the resistance to wear. Synthetic rubber, polybutadiene, SBR and natural rubber are used in tread.

Some modern tyre manufacturers use the "cap-and-base" tread construction. This technology allows the manufacturer to laminate two different compounds of rubber i.e. one on top of the other. The compound closest to the steel belt is the base rubber that is formulated to resist heat generation while the top compound is cap rubber, which is in contact with the road and it is softer to allow for a better grip. The tread design has been updated and the modern trend goes for unidirectional design that offers an excellent water drainage without sacrificing the dry handling facility and grip.

The other innovation is an asymmetrical tread pattern that has large ribs of rubber on the outside shoulder with smaller ribs on the inside shoulder. The larger ribs at the outside stiffen the tread, where the largest magnitude of lateral load is concentrated during cornering.

The factors like tyre noise, straight-line acceleration and cornering are also affected by the tread design and research and development of newer ideas is going on. The tread blocks on most of the tyres vary in size in an effort to break up the resonance developed when the blocks of equal size come into contact with the road surface.

3.4 *Mechanical Design*

Removing water quickly from the tyre surface is a key requirement in countries having significant rainfall. However, the water-removing proves may not displace water if the design follows aesthetics rather than mechanics. Circumferential grooves that are too wide can mean uneven tyre wear. Lateral grooves between the blocks also have a critical design angle whereas too great angle may lead to unacceptable tread wear and too narrow angle may develop excessive noise level. The larger the tread block, the greater would be the noise; while a uniform size would be noisier than blocks of varied size. However, too small blocks would have a negative effect on steering performance.

On many tyres, the surface area of tread blocks may vary upto 40%, while the layout on right and left parts of the tread are staggered to further reduce the noise. This spreads the noise-creating energy over a wider range of frequencies with lesser energy at any one single frequency. Continental Tyre Group has developed a bionic contour on its tyres, inspired by cat's paws. Under braking, the tread spreads out like a cat's paw enabling greater forces to be transmitted to the road.

3.5 *Computer Assistance*

The design features of tyre are first drawn on Computer Aided Design (CAD) software and then analysed for performance using Finite Element Analysis (FEA) software. Computer models of tyres can be analysed in terms of beams or plates. Beams represent forces that act in a line whereas plates can undergo complex flexing in three dimensions. Software helps tyre engineers balance the conflicting demands of tyre design without having to make the tyre actually and test the physical models. Anyway, selecting the best tyre design for any given car is subjective. Appearance and fashion are the key considerations.

Tyre diameter is often determined by appearance in relation to the vehicle as a whole, while the proportion of tyre width to sidewall height called "aspect ratio" is also subject to fashion. Selecting a tyre design for road holding and ride quality is also subjective. One method is to fit a previously used type to the new vehicle and assess it dynamically with a drive on a test track. This acts as a reference on which it would be necessary to base a set of changes.

Certain parameters can be given a measurable objective like ply rating, braking distance in wet and dry conditions, noise level off the road surface, frolling resistance, load rating and wear. However, subjective testing is needed to decide parameters like over and understeer.

3.6 *Belt Package*

The belt package consists of bracing cords. The belts are designed to help reinforce the carcass against the internal pressure created by inflation pressure, centrifugal forces of the tyre rotating at speed, shocks from bumps and potholes on the rough road and to enhance lateral stiffness to the treat.

4. Tyre Life and Causes of Tyre Wear

There are many factors that influences tyre life. Some of these are:

(i) Cuts: Cuts in the tyre permit damp to reach the casing plies. This is harmful for both textile and steel casing affecting safety and tyre life.

(ii) Contamination: Care must always be taken to ensure that contaminants like oil, paraffin and other chemicals donot come into contact with the tyres. It is highly dangerous to use possible rubber solvents like oils to assist tyre fitting. Contamination of tyres with substances e.g. hydraulic fluid, fuel oil or grease should be avoided as these materials have a deteriorating effect on rubber. Contaminants should be removed as quickly as possible preferably using a denatured alcohol or detergent and clean water.

(iii) Dual Tyre Matching: For tyres of the same type, size and ply rating, the amount of wear should be matched together.

(iv) Tyre rotation: Tyres are to be rotated only if the wear is irregular or if the wear rate is uneven.

(v) Vehicle Maintenance: This includes wheel alignment, wheel balancing, mechanical irregularities etc. as detailed below:

(a) *Misaligned Wheel*: This is the most common cause of irregular tyre wear. It is important that the tyre runs at the correct plane to prevent scrubbing of the tyre that causes rapid tyre wear. Wheel alignment must be undertaken regularly. A regular visual inspection would enable to spot any misalignment well before the outcome becomes costly. Some of the signs of misalignment are faster wear, visual scrubbing of rubber on the tread surface, uneven wear on one side of the tyre, 45°-wear pattern on tread, etc.

(b) *Wheel Balancing*: Tyre life will be affected adversely for any unbalance in the wheel assembly. The effect is seen in the form of an uneven wear pattern on the tyre tread. Static and dynamic balancing of the tyres must be checked regularly. An unbalance can be noticed early from the excessive vibration developed on the steering wheel of vehicle.

(c) *Mechanical Irregularities*: Many of the components of a vehicle may contribute to the development of irregular tyre wear. It is essential to inspect the tyre wear regularly to identify any problem that might exist. The following mechanical components can lead to developing any irregular tyre wear, grabbing or unequal brake clearances, spread axles, bent axles, sagging axles, twisted or cracked chassis, worn or loose wheel bearings, damaged or worn suspension, damaged or worn wheel assembly, worn bushes or ball joints, damaged or worn steering components etc.

(vi) Rolling Resistance: A major problem faced by the tyre designer's of heavy earth movers is to arrive at the compromise between low rolling resistance and wet grip. Rolling resistance is the amount of energy that a tyre absorbs as it revolves and deflects. The lower the rolling resistance, the lesser would be the fuel requirement to propel the vehicle forward. However, the lowering of rolling resistance would mean traditionally the reduction in wet grip performance, which is not acceptable. On the other hand, the rolling resistance is affected by low frequency distortion whenever there is any occurrence of deflection of the tyre as it revolves. It also requires the compounds that would absorb low quantities of energy like the low hysteresis compounds. This is the reason why it has not been possible to provide tyre rubber that reduces the rolling resistance and at the same time increases the wet grip.

(vii) Inflation Pressure: The relative demerits of inflation pressure and loading magnitude are :

(a) *Overinflation/Underloading:* This leads to the reduction in cushioning power of the tyre. The tyre is more susceptible to impact, penetration, abrasion and reduced road-tyre contact. Further, this affects the handling characteristics of the vehicle e.g. excessive center wear of tyre.

(b) *Underinflation/Overloading:* This is the major cause of tyre burst. Underinflation causes excessive flexion of the tyre sidewall. This leads to overheating and ultimately causes break-up and tread separation. Underinflation or overloading creates excessive stresses and heat that can lead to tyre failure. This may then result in vehicle damage and/or serious injury or death of operating personnel. A proper inflation would extend the tyre life and save the fuel also.

(viii) Heat: An underinflated or overloaded tyre flexes too much causing internal shear and eventual separation in the fabric. This mechanical effect is damaging enough, but the heat that generates is far more obstructive. Bending of any material may generate heat. Rubber is a poor conductor; so the heat developed by flexing builds up in the tyre. If it passes 120°C, the rubber will loose strength and some fibres will be weakened. At 160°C i.e. the temperature at which many tyres are cured and retreated, the plies may get separated and the tyre may blow out.

(ix) Incorrect Alignment: Alignment refers to the suspension angles on the vehicle. There are three main criteria leading to this kind of tyre wear i.e. toe, camber and castor.

(x) Grip: The surfaces of the road may provide different amount of grip. The compound of the tyre rubber determines the amount of interlock between the tread and the road surface. Under sliding, braking, wheel spin or break away condition, the interlocking rubber is actually ripped off the tyre and this leaves skid marks on the road. Tyre width, suspension settings, wheel-rim width, temperature, tyre pressure and load would influence the tyre grip. Grip is responsible for traction while accelerating or braking and it also generates cornering force. A tyre with high grip will corner fast, accelerate rapidly without spinning and brake quickly.

(xi) **Load:** This factor affects the flexion or deflection of the sidewall of tyre. As the speed of the vehicle increases, so does the frequency of sidewall deflection, which in turn causes an increase in heat that gets generated by the deflection. Tyre failure will almost always result if this is allowed to continue for too long a period.

(xii) **Abnormal Tyre Wear:** Mechanical irregularities such as incorrect or excessive suspension settings, incorrect wheel alignment, excessive wear in suspension components and faulty steering or braking system are likely to cause uneven or rapid tread wear. Of these factors, those causing incorrect toe angles are responsible for the most accelerated mechanical wear.

5. Damage and Wear of Tyres

(i) Tread Detachment

Overheating of the tyre often causes separation of the tread from the casing. Overheating is mainly caused due to driving with insufficient tyre inflation and the use of incorrect type of tyres.

(ii) Air Pockets

Infiltration of inflating air through the rubber is generally due to incorrect procedure adopted for tyre repair and perforation of airtight lining of a tubeless tyre caused by the misuse of tyre levers during fitting, accidental perforation in service or repair, buffing of the tubeless liner not being restored to its airtight state.

These perforations may allow the air to penetrate upto the casing cords, which may channel it to the ends of the turn-ups of the plies. The air accumulates and causes separation, which is characterised by small blisters or cracks.

Once the air becomes trapped in a tyre, the temperature of trapped air may increase when the tyre temperature increases under normal service conditions. This heated trapped air expands and gradually forces the separation thus allowing more air into the cavity and this air will also expand to cause further separation.

(iii) Sidewall Cuts and Rupture of Sidewall

This may not necessarily have any immediate consequence; but it could gradually extend and develop into a rupture of the sidewall. It is necessary to have all cuts repaired and avoid worsening of the damage.

The rupture of sidewall may be caused by underinflation, overloading, air pockets or an impact.

(iv) Rupture Due to Tread Cut

A cut in the tread that may or may not have affected the crown or casing plies would weaken that area of the tyre. The penetration of grit and moisture into the cut will gradually extend the damage as will the flexing of tyre in service. This is likely to develop rupture of the plies. This rupture may not occur until long after the cut and can cause sudden deflation. A cut in the tread is not always seen easily, but this can be sometimes detected from excessive wear at the damaged area.

(v) Impact Ruptures

When a tyre strikes an obstacle, the casing becomes badly distorted and this may lead to a rupture of the plies. The effects of the shock may not be noticed immediately and the tyre may not be marked or bruised. More often the immediate effect is localised rupture of only some plies, possibly without external evidence. Repeated flexion of the casing gradually extends the damage that will eventually appear externally.

(vi) Pinch Rupture

In such cases the tyre gets pinched between the obstruction on road and the rim. This can result in rupture of the plies without any immediate visible damage. This damage could occur later in service, showing itself as a sidewall bulge or even a sudden deflation.

(vii) Bead Damage

The damage of bead is usually due to overloading, underinflation or incorrect fitting. These factors impose excessive strain that can lead to splitting and separation in the bead area. Such damage may be in the form of unwrapping of the casing ply from around the bead core and rupture of the casing in the bead area. Similar damage can also be caused due to excessive heat from brake drums. Excessive flexing of an incorrectly fitted, overloaded or underinflated tyre can cause the beads to chafe against the rim flange.

(a) *Incorrect Fitting or Removal*

Incorrect use or positioning of tyre levers unsuitable for the tyre, may cause damage in the area of the bead toe in particular. This in turn would cause air leaks in the casing of a tubeless tyre or damage the tube in a tube type tyre.

(b) *Overinflation*

Excessive inflation pressure i.e. just like fitting a tyre on the incorrect size of rim can also cause the bead core to break.

Various types of tyre damages are shown in Figure 1.

(a) Cut damage (b) Dry braking flats (c) Tyre burst

(d) Excessive brake heat (e) Trade groove cracking

Fig. 1: Different Types of Wounds in Tyres

6. Tyres Testing Techniques

A number of experimental methods have been used to improve the characterisation of tyre wear and tyre wear particles as given below:

(i) The chemical analysis of bulk tyre material has been done to determine concentration of rubbers, metals and a wide range of other substances. Sample collection of tyre debris *in situ* and GC/MS (gas chromatography/mass spectrometry) has been used to characterise tyre wear particles.

(ii) The sampling of airborne particles near to roads and receptor modelling techniques have been used to determine the emission factors for tyre wear. Tracer techniques are combined with the receptor modelling method to find the contribution of tyre debris in the ambient samples.

(iii) Gravimetrical definition of tyre wear using test vehicles (i.e. by weighing of their tyres after a predefined period of use) is used to determine the loss of tyre material in mg/tyre/km.

(iv) Laboratory-based tests to simulate real-world tyre wear.

Using the techniques mentioned above, the wear rate of tyre is found to be in range of 8 mg/tyre/km to 90 mg/tyre/km. The chemical analysis of tyre tread reveals an abundance of Zn for tyre material and lower concentration of other metals, such as Ca, Fe, Pd etc. The bulk of tyre tread is also composed of a variety of rubbers like natural rubber, SBR (Styrene Butadiene Rubber), butadiene rubber etc.

The chemical composition of tyre wear particles is found to include quantities of EC (Elemental Carbon), OC (Organic Carbon), PAH (Polycyclic Aromatic Hydrocarbon) and several metals e.g. Zn, S, Si, Ca, Fe etc. Further, there is a conflict about the size distribution of tyre wears particles. Some researchers indicate that tyre wear particles are only in the coarse-mode fraction (2.5-10 μm), while other surveys found that they are ranged between 0.01 and 30 μm.

The various tests on material properties as are routinely carried out for tyre wear by the chemical testing laboratories include ash content, density, melting point, solvent extractables, boiling point, extractable proteins, pH of liquids, total solids, composition analysis, fluid viscosity, polymer blend ratio, volatile matter content, corrosion resistance, metallic ion content and softening point.

6.1 *Non-Destructive Testing*

Holographic Non-Destructive Testing Techniques (HNDT) are used to locate and evaluate cracks, disbonds, voids, delaminations, inhomogeneity and residual stresses in a test sample without destruction of the sample. The holographic interferometry techniques are applied for non-destructive testing of materials.

The object can be stressed by mechanical stressing, pressure or vacuum stressing, thermal stressing, vibrational stressing and magnetic stressing. The stressing of the object can create gross deformation and rigid body motion of the object. This will produce fine interference fringes in the interferogram if the test area is large. In such a situation, the interference fringes around the flaw will be very fine and it would not be detected by unaided eye. By using fringe control methods, the effects of gross deformation and rigid body motion can be compensated.

The rubber products testing unit covers interalia the following:

- Retread manufacture and troubleshooting.
- Tyre structural endurance test.
- Load deflection and footprint measurement.
- Determining tyre forces at slip/camber angle.
- Dynamic circumference measurement.
- Rolling resistance measurement.
- Bead unseating force.
- Plunger energy measurement.

7. Improved Wear Resistant Tyres

Several researches have been done to improve the tyre quality in automobiles as well as in mining vehicles due to the rock-bed over the haulage road surface. Some of them are described in brief on next page.

7.1 *Tyre Protection Chain*

The tyre protection chains are used to protect the tyres of wheeled loaders, excavators, dump trucks and graders operating in quarries, opencast and underground mines, tunnels, cement plants, steel mills, forestry etc. The pictorial view of tyre protection chain is shown in Figure 2.

Fig. 2: Tyre Protection Chain (TPC).

7.2 *Tubeless Tyres*

It has been over a century since the time Dunlop patented the 'mummified wheel' to the modern radial tyres of today. Despite all the improvements a tyre has undergone, one thing remained unchanged i.e. when it is inflated to the optimised level and that inflation is kept constant so that it can deliver maximum comfort and performance. This is one of the basic reasons for which all tyre manufacturers try to focus on during the development stage of a tyre, which can have the best air retention ability. Usage of a tube or an extra air container within the tyre was regarded as the best solution for many years.

This concept was not pursued until late 1939 when the requirement for the first amphibious tyre was felt. The 120 × 33.5 – 66 smooth tread Marsh Buggy tyres, by far the largest tyres produced then, were used on Admiral Byrd's Snow Cruiser. This vehicle was capable of carrying very heavy loads over all sorts of terrain, even float on water. These were off-the-road tyres, flexible but inextensible pressure vessel that were pre-stressed and skin-stressed by air pressure. To produce such tyres Goodyear at Akron employed the idea of Litchfield, using nylon cords for the first time and a new synthetic rubber compound called 'Chemigum' was developed to line the inner casing of this tyre and to lighten its weight eliminating the tube.

This enhances high-speed performance while achieving a better cornering ability on the tubeless. The other advantages are the absence of a tube making the tyre lighter in weight, thus has lesser chance of vibrations, which means that it leads to a better fuel saving. Even the rolling resistance in a tubeless radial is lower when compared to a tube type radial tyre. This is due to the fact that the tubeless tyre sidewall is more supple as there is no internal body to create a friction. This also helps the tyre to run cooler as it eliminates heat generation caused by the internal shuffling of the tube.

7.3 *Retread Tyres*

The only part of a tyre to wear in normal use is the tread and it makes economical sense to give a new lease of life to an otherwise good tyre by what is now generally known as re-treading.

8. Conclusions

Tyre technology is a very specialised technology and a tyre designer is faced with the magical task of trying to satisfy all the needs of the vehicle manufacturer and user. Much development work is carried out by the tyre manufacturers into their tyre compounds but it can often be a hit-and-miss affair. This is why the recipes for tyre compounds are such closely guarded secrets. The most important factors in tyre care are:

- Proper inflation pressure
- Proper vehicle loading
- Proper tyre wear
- Regular inspection
- Good driving habits
- Vehicle condition.

At present research work is going on in different countries all around the world mainly to improve the durability and wear resistance properties of tyres in mining use. Many of them have developed a few different types of wear resistant tyre compound, which are likely to find their commercial application in future.

9. References

[1] Fauser, P. Particulate air pollution with emphasis on traffic generated aerosols, Risp National Laboratory, Denmark, 1999.

[2] Hildemann, L.M., Markowski, G.R. and Cass, G.R., Chemical Composition of Emissions from Urban Sources of Fine Organic Aerosol, Environmental Science and Technology, 25(4), p. 744–759, 1991.

[3] Rogge, W.F., Hildemann, L.M., Mazurek, M.A., Cass, G.R., Sources of fine organic aerosol road dust, tyre debris and organometallic brake lining

dust: Roads as sources and sinks. Environmental Science and Technology, 27(9), p. 1892–1904, 1993.

[4] Emission factors for fine dust from road traffic, Second International ETH Workshop on Nano-particulate Measurment, August, 1998.

[5] Hewit, C. and Rashed, M.B., An integrated budget for selected pollutants for a major rural highway. The Science of Total Environment, 93, p. 375–384, 1990.

[6] James, D.I. (ed.), Wear of Rubber, London, U.K, 1967.

[7] Premomoy Ghosh, Polymer Science and Technology of Plastics and Rubber, p. 383-387, 1992.

[8] Kumata, H., Hideshige, T. and Ogura, N., Determination of 2-(4-Morpholinyl) benzothiazole in environmental samples by a gas chromatograph equipped with a flame photometric detector, Analytical Chemistry, 68, p. 1976–1981, 1996.

[9] Pal, Samir K., Misra, B., Bhowmick, A.K. High speed wear of conveyor belt rubber by various rocks, Journal of Plastics, Rubber and Composites Processing and Applications, Vol. 25, No.7, p. 332–339, 1996.

[10] Fishman, R.L. and Turner, J.R., Tyre wear contributions to ambient particulate matter, Air Quality Laboratory Report No. WUAQL-0399-03, Washington University, St. Louis, USA, 1999.

[11] James, E. Mark, Erman, Burak, Fredrick R. Eirich, Science and Technology of rubber, London, U.K, 1968.

[12] Pal, Samir K., Bhowmick, Anil K. Design of Rubber for Wear Resistant applications, Rubber News, Indian Institute of Technology, Kharagpur, p. 35-38, Sept. 1989.

Thermogravimetric Studies of Different Tyres and Tubes

Emranuzzaman, M.K. Choudhary, Jagdish, B.M.P. Pingua and Nabiullah

Central Mining Research Institute, Dhanbad-826001

1. Introduction

John Boyd Dunlop invented the rubber tyre during the late nineteenth century i.e. in the year of 1888. The pneumatic tyre was actually invented and patented in the year of 1845 by R.W. Thompson. The concept of R.W. Thompson was more advanced as compared to that of Dunlop. Rather than just using a rubber tube filled with air, Thompson packed a number of thin inflated tubes inside a leather cover. The pressures could be varied for different ride conditions and, of course, one puncture would not be terminal. For no obvious reason (except possibly the lack of roads!) these tyres could never catch the eyes; on the contrary the solid rubber tyres mounted on various vehicles starting from bicycles to steam traction engines came in vogue. In 1884 the idea of leaving a hole through the centre of the rubber was developed mainly to give a 'cushioning' effect. However, until December 19, 1888, when the first advertisement for a 'Dunlop Pneumatic Tyre' appeared in the "Irish Cyclist." On May 18, 1889, a cyclist using these tyres won a race in Belfast. According to the rough estimate that depends on molecular weight and structure of the rubber used in car tyres, the tyres are made of 28% carbon black, 27% synthetic polymer, 14% natural rubber, 10% wire, 10% oil and 11% other materials.

Elastomeric components (i.e. rubber parts) are very often the items limiting the service life of today's machines, vehicles and appliances. The requirements of new polymers and improved formulations generated the need to sort out the serviceables from unserviceables. The actual sorting and ranking procedure varies with regard to inherent properties of the polymer and the intended use of the final product. The rubber material in concept normally means elastic, extensible and flexible mass. Further, the rubber materials can get deformed hundred percent under loading and return rapidly to their original dimensions when the load is removed. Materials with rubber-like properties at room temperature are generally called elastomers. Elastomers that can be cross-linked with vulcanizing agents are called rubbers. Rubber is a macromolecular substance with a glass transition temperature lower than 0°C, composed of long flexible molecules, randomly agglomerated and entangled in the chains producing mechanical knots. The macromolecules are formed by the repetition of one or more alternating monomers. For example, natural rubber is formed by the repetition of the isoprene unit. On the other hand, the styrene-budadiene copolymer rubber is composed of monomeric units of butadiene and styrene repeated in various ratios and sequences.

Generally, the various types of rubber would differ from each other as below:

- In the chemical nature of the monomer units
- In their mutual bonding and arrangement
- In molecular weight, molecular weight distribution, branching level, presence of gel, etc.

The automobile tyre is manufactured as *vulcanized* rubber. The natural rubber is additional polymer of *cis-2*-methyl-1, 3-butadiene (isoprene).

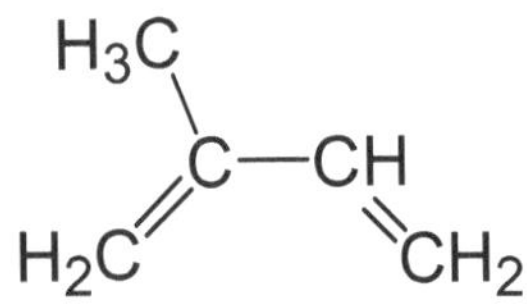

As a rule of thumb, the scrap tyre generation in industrialized countries is approximately one passenger car tyre equivalent (PTE, 20 lbs., 9 kg) per population in a year. It is estimated that 2-3 billion scrap tyres are stockpiled in illegal or abandoned piles throughout the U.S. This figure would represent the cumulative scrap tyre generation over a period of ten years. For EU member states, illegal or semi legal scrap tyre piles develop in the same order of magnitude. Obviously, the hazards associated with the uncontrolled disposal and accumulation of huge amount of tyres at the outdoor would have the potential for large fires that might be extremely detrimental to the environment. Also, it has been found that air and soil pollution becomes even worse if attempts are made to extinguish the fire with foam or water. For this reason, scrap tyre fires are often allowed to burn out in a controlled manner until the entire pile is exhausted. The second problem often caused by illegal tyre piles would lead to development of some diseases carried by mosquitoes, which tend to breed in a great number in the protected water puddles of the tyres. Especially, in the areas of warmer climates, mosquito-borne diseases like encephalitis and dengue fever have been reported around large tyre piles.

Guillermo San Miguel, *et al.* have studied the project on active carbon production from tyre rubber production, characterization and application. The context of this study was the problem of appropriate disposal or reuse of scrap tyres, of which the UK produces about 500000 tonnes per annum. At present, about 50% of these tyres are stockpiled or land-filled. Amongst the alternatives are possible production of activated carbons from tyres by thermal treatment as well as by modifying and adopting the carbonization and activation techniques used for the production of commercial activated carbons. Further, this study evaluated the activation techniques using carbon dioxide or air.

Yuan Zhang, *et al.* have described the effects of tyre properties and its interaction with ground and suspension system. Also the dynamic behaviour of the vehicle is very important to both the tyre manufacturers and the automakers, specially the suspension designers. Due to the criticalities involved in this connection e.g. complex material properties, nonlinearities, complex geometry, etc., various assumptions and simplifications are required to deal with such problem theoretically.

Kennedy, R.H. *et al.* made a study on the tyre temperature prediction. During the postcure inflation, the temperature distribution was predicted as the tyre got cooled. A finite element based transient heat transfer analysis was done in this process considering the appropriate material conduction and convection coefficients as determined by laboratory experiments. The initial temperatures to be applied to the tyre model were determined from the measurements on a range of tyre sizes. Measurements were made during as well as at the end of the postcure inflation cycle for making a comparison with the predicted values.

Although there are many methods of ranking, the most accurate results obtained through the use of precision instruments are used to evaluate a specific feature to render the proper and accurate data. The ability of an elastomeric material to withstand the high temperatures can be determined by measuring the changes for different properties, including tensile strength, modulus of elasticity, elongation, hardness, weight, resilience, shape, volume, tear strength, chemical composition, color and surface features. However, many of these changes are interdependent and are of no consequence or they do not provide consistent information.

The most consistent and meaningful behaviour of elastomer at the degradation temperature is weight loss. Therefore, a means of detecting the onset weight loss versus temperature would provide a single temperature value for each material, i.e., 140°C for poly (ethylene-co-propylene) elastomer in air. While several complicating effects interfere with the credibility of the onset temperature as a measure of the upper use limit, it is a reasonable method for ranking new materials for service in high temperature environment.

Thermogravimetry (TG) can be used to rank materials in order of increasing service capability in high temperature environment. TG is particularly well suited for evaluating elastomers because the predominant mechanism of thermal degradation would involve weight losses. The temperature at the onset of weight loss has been a very useful figure, indicating the upper limit of high temperature capability. The data obtained would depend on reliable instruments, behaviour of instruments, careful preparation of materials, knowledge of sample history, purity of purge gases, etc. Interpretation of a TG curve for the onset temperature requires an insight into elastomer composition, mechanism of degradation and careful scrutiny of the complete weight loss curve. A valuable ranking of new materials as well as an upper thermal limit can be obtained by TG, although the realistic long-term service temperature will be lower due to the added features of the actual environment.

2. Experimental Procedure

The scrap tyre and tube of bicycle and scooter of different companies were taken and analyzed by Metller, T.G. The tyres and tubes of the bicycles were of Metro & Co. and that of the scooter were of MRF Ltd. The starting temperature was 25°C and the end temperature was 800°C. The rate was 20 K/min.

3. Results and Discussions

Elastomers refer to a composite of elastomeric polymer, crosslinked into a network with reinforcing fillers, crosslinking agent residues, inhibitors, plasticizers, antioxidants etc. Serviceability of the same depends on the retention of mechanical properties. Such properties would depend on the three dimensional elastomeric network.

Three general mechanisms describe most of the observed modes of polymer degradation at high temperatures. The most frequently encountered factor is the loss of small molecules followed by charring, while the other is the frequency of occurrence that appears to undergo breakdown into volatile fragments leaving no residue and often preceded by the loss of small molecules.

Another mechanism is unzipping that regenerates the monomer from which the polymer is made. In such mechanism some specialities are also available where several monomer units or a specific molecule is released from the polymer chain. A good indicator of degradation mechanism is the shape of the TG curve. The gradually increasing rate of weight loss indicates the loss of small molecules. The change from small molecule loss to char is often gradual, but the slope change indicates the slower process for small molecule release from char. Most often, the elimination of fragments begins through the release of small molecules. The transition to fragmentation depends on the polymer system. The more liable structures are eliminated at first and possibly at a high rate, thereafter the more stable structure is broken down at the higher temperature and usually at slower rate.

Complete volatilization of the polymer part of an elastomer below 600°C is a good indication that the carbon char was not formed. The onset of rapid weight loss with no change in rate until the polymer part is volatilized indicates unzipping. The monomer is usually evolved but may react before escaping the hot crucible.

Table 1 shows the weight loss temperatures for various elastomers.

Table 1: Onset of Weight Loss Temperatures for Various Elastomers

Sl. No.	*Material*	*Thermal Stability (Helium)*	*°C Air*
1.	Poly (chlorobutadiene) gum	195	---
2.	Poly (chlorobutadiene) elastomer	170	---
3.	Poly (vinylidenefluoride-co-hexafluoropropylene) elastomer	240	---
4.	Poly (fluoroalkoxy phosphazene) elastomer	220	210
5.	Poly (ethelene co-propylene) elastomer	---	140
6.	Poly (trifluoropropyl, methyl siloxane) elastomer	250	220
7.	Poly (trifluoropropyl, methyl siloxane) fuel tank sealant	---	220
8.	Fluorosilicon-fluorocarbon hybrid elastomer	275	130
9.	Poly (tetrafluoroethelene-co-perfluoromethlenevinylether) elastomer	280	---
10.	Poly (tetrafluoroethelene-co-Trifluoronitrosomemethane) elastomer	200	---
11.	Dimethyl siloxane carborane elastomer	300	290
12.	Poly (perfluoroalylene perfluoroalkyl triazine) elastomer	310	---

Table 2 to 5 show the weight loss temperature for various tyres and tubes. The result shows that cycle and scooter tyres and tubes have lost their weight from 29°C. This may indicate that the scrap tyre and tube have lost their first polymeric compounds weight at that temperature. However, the end temperature of first polymeric compound in scooter tyre was 145°C whereas that for the cycle tyre was 441°C. This may indicate that scooter tyre moves faster than the cycle tyres. Hence, the heating effect on scooter tyres is more than that for the cycle tyres and the polymer in scrap tyres may loose their polymeric chain. The end temperature of first polymeric compounds in scooter tube was 125°C whereas that for cycle tube was 181°C. Figure 1 to 4 show the thermogravimetry curves of scooter and cycle tube and tyre.

Table 2: Weight Loss Temperature of Scooter Tyre

Sl. No.	*Starting Temperature (°C)*	*Peak Temperature (°C)*	*End Temperature (°C)*	*Percentage (%)*
1.	29	49	145	–2.9262
2.	145	301	381	–8.7787
3.	381	453	493	–45.292
4.	493	561	685	–45.674

Table 3: Weight Loss Temperature of Scooter Tube

Sl. No.	*Starting Temperature (°C)*	*Peak Temperature (°C)*	*End Temperature (°C)*	*Percentage (%)*
1.	29	49	125	–3.8330
2.	125	409	489	–60.917
3.	489	557	697	40.657

Table 4: Weight Loss Temperature of Cycle Tyre

Sl. No.	*Starting Temperature (°C)*	*Peak Temperature (°C)*	*End Temperature (°C)*	*Percentage (%)*
1.	29	377	441	–48.038
2.	441	553	657	–27.454

Table 5: Weight Loss Temperature of Cycle Tube

Sl. No.	*Starting Temperature (°C)*	*Peak Temperature (°C)*	*End Temperature (°C)*	*Percentage (%)*
1.	29	49	181	–3.4590
2.	181	385	429	–48.086
3.	429	453	561	–10.604
4.	561	641	681	–7.0031
5.	681	717	757	–3.8560

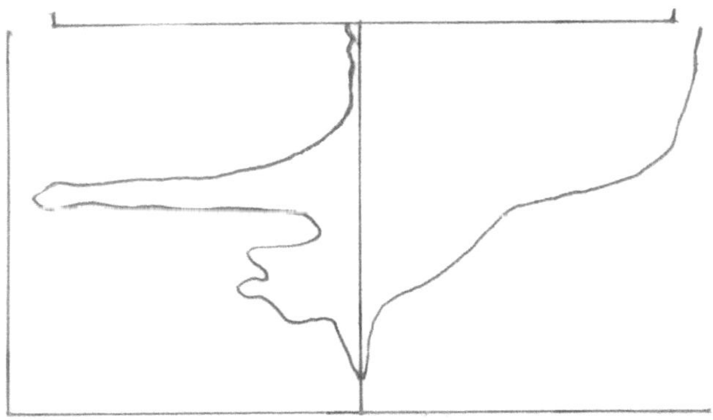

Fig. 1: TG curve of cycle tyre

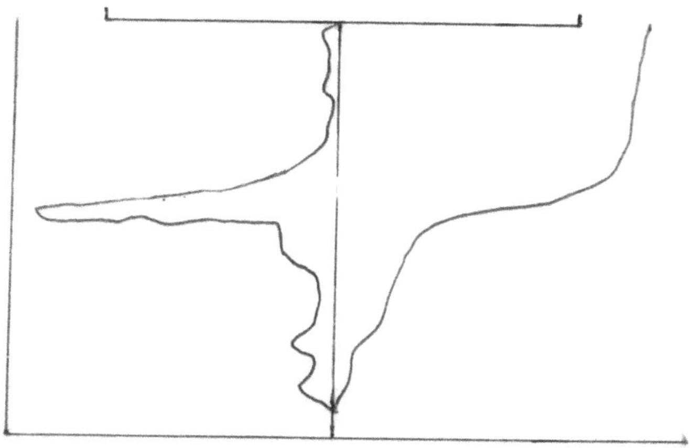

Fig. 2: TG curve of cycle tube

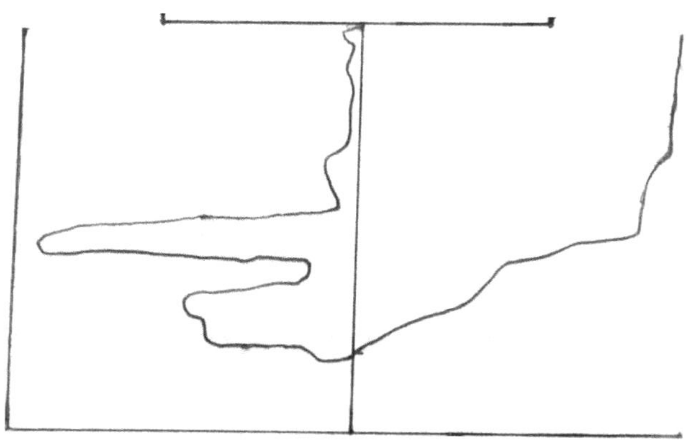

Fig. 3: TG curve of scooter tube

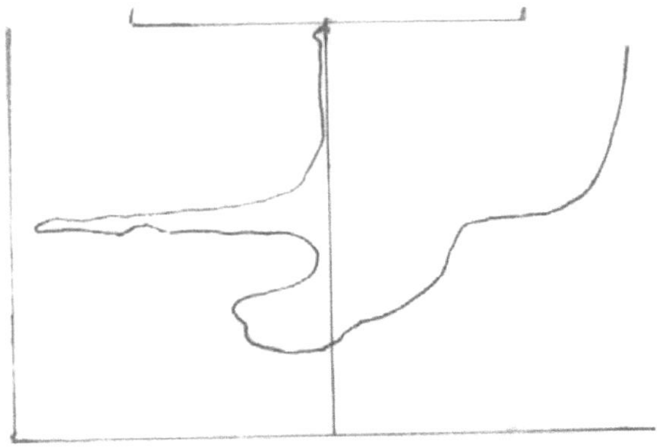

Fig. 4: TG curve of scooter tyre

4. Conclusions

(i) The automobile tyre is the *vulcanized* rubber. The natural rubber is the additional polymer of *cis*-2-methyl-1, 3-butadiene (isotope).

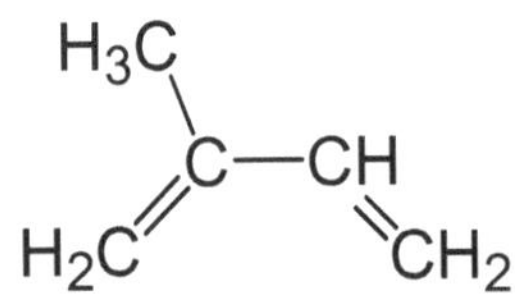

(ii) In warmer climates mosquito-borne diseases like encephalitis and dengue fever have been developed around large tyre piles.

(iii) The starting temperature in cycle and scooter tyre and tube was 29°C.

(iv) The end temperature of first polymeric compounds in scooter tyre was 145°C whereas that for cycle tyre was 441°C. This may indicate that the scooter tyre moves faster than the cycle tyre. Hence, the heating effect on scooter tyres is more pronounced than that on the cycle tyres and the polymer in scrap tyres may loose their polymeric chain.

(v) The end temperature of the first polymeric compounds in scooter tube was 125°C whereas that for the cycle tube was 181°C.

5. References

[1] http://www.tut.fi/plastics/tyreschool/moduulit/moduuli_6/hypertext/1/1_1.html dated 18/11/2003.

[2] http://home.snafu.de/kurtr/str/en_sr.html dated 18/11/2003

[3] San Miguel, G., Fowler, G.D., Sollars, C.J., "Preparation of active carbon from tyre rubber by carbon dioxide activation", *Proc. European Carbon Conf,* Newcastle upon Tyne, 163–165, 1996.

[4] Zhang Yuan, Hazard Curtis, "The Effects of Tire Properties and their Interaction with the Ground and Suspension on Vehicle Dynamic Behaviors—A Finite Element Approach" Abstract on-line Databse-conference. www. tire society.org.

[5] Kennedy R.H., McMinn, M.S., "Tire Temperature Prediction During Post Cure Inflation," Abstract on-line Databse-conference. www. tire society.org.

[6] Paciorek, K.L., and Kratzer, R.H., and Kaufman, J., "Oxidative thermal degradation of polytetrafluoroethylene," J.Polymer Sci. (Polymer chemistry edition), Vol. II, 1466–1474, 197 3.

[7] Conley, R.T., "Thermal stability of polymers," Marcel Dekker, Inc., New York, Vol. 1, 1970.

[8] Griffin, W.R., "Themogravenetry of Elastomers" Thermal Analysis: Comparative studies on material, edited by Kambe, H., Garm, P.D. John Willey & Sons, pp.142, 1974.

Oxidative Degradation of Tyres

Tanushree Chowdhury and N.M. Mishra

Department of Applied Chemistry, Indian School of Mines
Dhanbad-826004

1. Introduction

Disposal of waste tyres is a global problem and this has not yet been solved anywhere. Tyres are manufactured from petrochemical feedstocks such as styrene and butadiene, both of which are being classified as human carcinogens. Styrene is a benzene derivative and burning of tyres releases styrene and several benzene compounds. Butadiene is a highly carcinogenic 4-C compound that may be released from the Styrene-Butadiene (SBR rubber) polymer form during combustion.

Aromatic extender oils comprise about 25% of most tyres and are known to cause cancer in laboratory animals as well as being suspected as human carcinogens. These are highly aromatic multiple benzene containing chemicals petroleum waste materials with complex ring structures that are even more difficult to burn than benzene, which has a highly stable ring structure. The thick black oil and black smoke that emit out during tyre burning are due to aromatic extender oils.

In this paper an alternative way of disposal of scrap tyres is suggested. Chemical oxidative degradation of tyres has been carried out and the resulting product is characterized using TGA and FTIR. The product that is formed is carboxylic acid and it can be used as a starting material for various reactions.

2. Materials and Methods

The materials used in the experimentation process are:

(a) Pieces of scrap tyres

(b) 0.1 M acidified $K_2Cr_2O_7$ solution

Pieces of scrap tyres were exposed to UV radiation and TGA of these treated and untreated tyres were taken.

Acidified $K_2Cr_2O_7$ solution was added to some pieces of scrap tyres and the same was subjected to heating. The resulting product formed was characterized using FTIR. The chemical reaction involved in such process is given on next page:

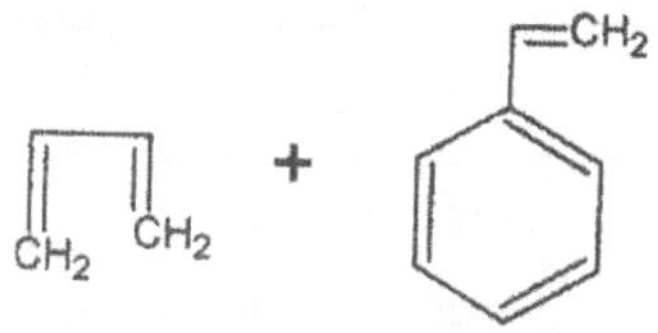

Butadiene Styrene

Carboxylic acid

$\longrightarrow$

$K_2Cr_2O_7$ Solution — Carboxylic acid

The TGA graphs of the treated and untreated tyres are presented in Fig. 1 & 2 respectively.

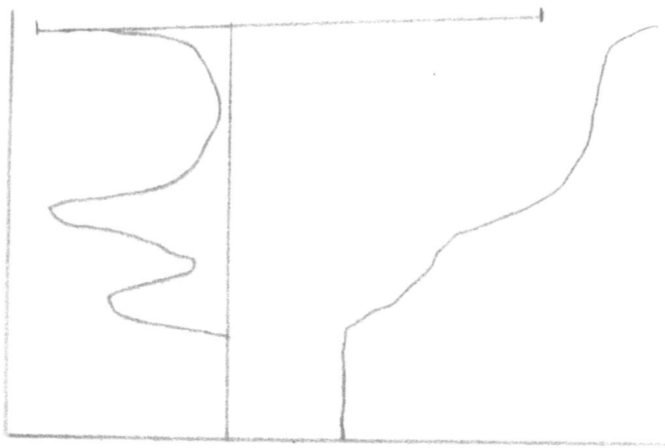

Fig. 1: TGA of Untreated Tyre

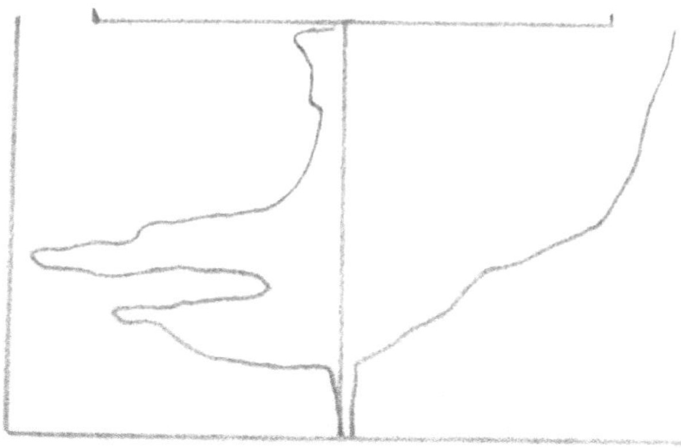

Fig. 2: TGA of Treated Tyre (tyre exposed in UV radiation)

Further, the FTIR spectra for untreated and treated tyres are given in Fig. 3 & 4 respectively.

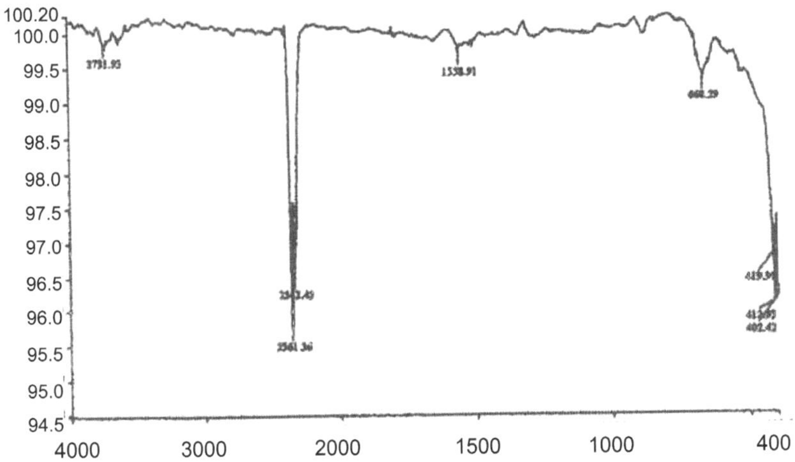

Fig. 3: FTIR Spectra Using Untreated Tyre

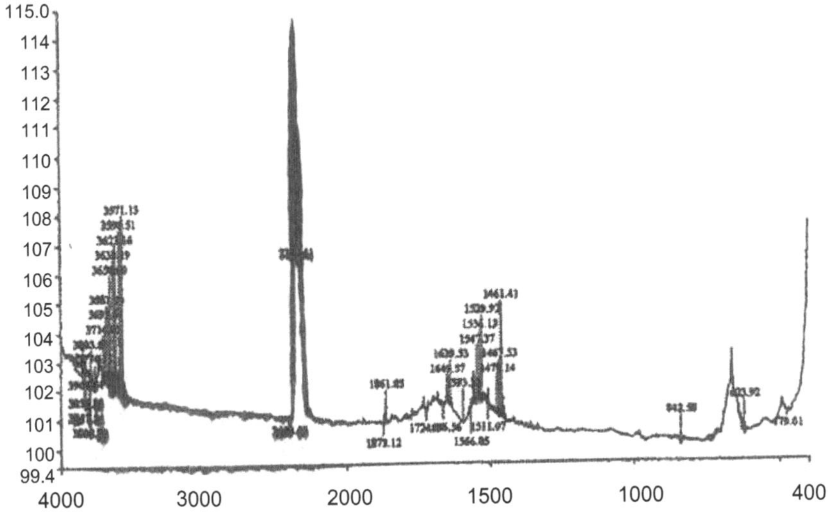

Peak at 1723 cm^{-1} shows that carboxylic acid may have been formed.

Fig. 4: FTIR Spectra Using Treated Tyre (with Acidified $K_2Cr_2O_7$ Solution)

3. Uses of Carboxylic Acid

It is the starting material for a wide range of derivatives. These include esters, amides, imidazolines, amines, salts that are subsequently incorporated into a variety of industrial applications. The specific uses of carboxylic acid are listed below:

- Adhesion and compatibility promoters in plastics and adhesives
- Pulp and paper process defoamers
- Lubricants in textile and latex processing applications
- Protector and gloss enhancer in polishes
- Adhesives for toners
- Intermediates in the manufacture of esters and resins.

4. Conclusions

- The oxidative degradation technique used in this paper is an alternative approach to burning of tyres that poses many environmental hazards.
- The end product that is formed in this method is carboxylic acid, which is characterized by FTIR and thus it can be a starting material for a variety of industrial reactions.

Disposal of Used OTR Tyres

M.S. Waldia[1] and S. Banerjee[2]

[1]Unit Rig, Nigahi Office, Terex Mining Corporation, Sidhi-486884
[2]Nigahi Opencast Project, Northern Coalfields Ltd., Sidhi-486884

1. Introduction

Since the late eighties, the disposal of used OTR tires has become a serious environmental concern in construction and mining industries world wide. The advanced countries of the world having a large population of automobiles, the problem of disposal of used OTR tyres was only a corollary to contemporary situation. This has led to evolving the mechanism e.g. "regulations and machinery for disposal of automotives tires".

The scrap generation from the used tires is as high as 1.2 billion per year around the globe. USA tops the list with more than 250 million scrap tyres per year. Currently, USA has nearly three billion scrap tyres in landfills and stockpiles scattered all over its territory. At one site near Modesto, California about 42 million tyres are stockpiled. The stockpile named "Mount Goodyear", is still climbing upwards at the rate of 20,000 tyres a day. The federal and state level committees/commissions in USA have predicted the current annual generation of scrap tyres as one per person. In the Indian context, the scrap tyre number is not available officially. However, the compilation of production figures of the manufacturers suggests an approximate figure for scrap tyres to be about 43.5 million (Year 2001–2002).

It is felt that there is considerable potential for reduced generation of scrap tyres in India if the following are taken care of:

- Use of outdated technology in the manufacturing of tyres.
- Use of bulkier biased tyres which has much lesser life as compared to radial tyres.
- Poor road conditions.
- Poor driving habits.

The Indian scenario is very much dismal because the Indian automotive, construction and mining industries are yet to start addressing the issue of scrap tyre disposal and it will not be off the mark to say that serious efforts are yet to be put in for addressing this issue. The efforts that are required in this connection are:

- Legislative regulations for necessary enforcement.
- Efficient mechanism for the processing of used OTR tyres.

OTR tyres are made of synthetic and natural rubber, nylon or cotton threads, steel cords and steel wires. The process of vulcanization is used for fusing the above mentioned parts. As a rule of thumb, an average highway truck tyre requires 100 litres of petroleum oil and almost 70% of this oil goes into carcass or the casing of tyre.

The consumption of OTR tyres in India is about 15000–20000 units per year. The increase in thermal power generation according to provisions laid in 10th five year plan will require the opening of new coal mines and capacity augmentation for the existing coal mines. This will exponentially increase the consumption of tyres by the year 2012, which is the target year for the Government of India's policy "Power on Demand by year 2012".

An effort has been made in this study to describe the methods used for tyre disposal and to make recommendations for possible reduction in the generation of OTR tyre scrap.

The following Table shows the energy required/released in various processes:

Energy needed to manufacture a tyre	32.0 kWh/kg
Energy needed to produce rubber compound of tyre	25.0 kWh/kg
Thermal energy released when incinerating scrap tyres	9.0 kWh/kg
Energy consumed in the process of grinding scrap tyres into crumb rubber (0.5 to 1.5 mm)	1.2 kWh/kg

Specific energy values, Source: W. Dierks Incorporating the Use of Recycled Rubber, Robert Snyder: Scrap tire disposal and reuse, compilation by Kurt Reschner.

2. Methodology for Disposal of Tyres

a. *Land Filling*

During the late seventies and onwards the OTR tyres were buried as landfill in land reclamation projects at USA. However, the reasons cited below discouraged the disposal of waste tyres in landfilling.

- Buried tyres were not compactable fully.
- The gases coming out of buried tyres used to cause ballooning effect.
- The tyres although made of organic substances were not bio-degradable.

Land filling in USA was restricted in 1991 (few states in USA still allow tyres of sizes below 18″ for landfilling) and will be completely banned from year 2006.

b. *Burning of Tyres*

Uncontrolled burning of tyres has proven to be very dangerous because the effluent gases, like benzene or its derivatives are released during the burning of vulcanized rubber and nylon chords. Several studies have already established benzene as carcinogenic toxic agent responsible for most of the respiratory diseases. In addition to polluting the atmosphere, the residues are known to damage the ground water resources and soil.

In Indian context, the legislation with the bite to ban the burning of scrap tyres is long overdue so as to fight the threat to environment by toxic pollutants produced by burning tyres in junkyards.

c. *Peeling of Tyre Plies into Layers*

In third world countries, the availability of cheap manpower has permitted this mode of disposal by taking out the reusable peels from tyres. The peels are used to manufacture low torque transmission belts, conveyor belts, insulating material, soles for foot wear etc.

National Solid Waste Association of India has estimated the volume of such business in India at about USD 375 million with the involvement of 250000 (approx) people engaged in the trade at different villages and towns in the country. However, the procedure is time consuming and cannot match the pace of ever increasing piles of used OTR tyres in this country.

d. *Devulcanizing of Tyres*

The recycling of cured rubber has proven to be really challenging. This is due to the reason that once vulcanized, the rubber cannot be easily melted and reformed into other products. There are techniques that can be applied for such purpose namely uncontrolled incineration as well as controlled incineration.

The controlled incineration/pyrolysis salvages rubber by melting the waste tyres under high temperature without air or oxygen. The process produces tar, iron residue, gases, rubber residue and water.

However, the incineration processes have potential for causing pollution.

e. *Shredding or Crumbing of Tyres*

Shredding or crumbing of tyres has provided the means to recycle the used OTR tires purposefully. It may be cited that 75% of waste tyres generated annually in US are recovered and recycled. The estimated US figure is in excess of 250 million waste tyres every year.

With the advent of modern technology, it is now possible to have an array of complex yet simple machines that can pulverize and separate all the ingredients of tyres effectively.

The advantages of reclaiming or recycling the rubber are as below:

- Recovered rubber costs half that of natural or synthetic rubber.
- Recovered rubber can have better properties than virgin rubber.
- Total energy requirement for reclamation is lesser than the energy required for production of virgin material.

- The crumbs can be used for manufacturing a multitude of rubber based products.
- Employment generation in such process is also another benefit.
- The energy produced by Tyre Derivative Fuel (TDF) is useful in power generation.

The shredded rubber has direct applications as below:

- The application of TDF under controlled burning can be seen in cement plant kilns and in power stations. In USA, approximately 75% of the recovered tyres are converted to tyre-derived fuel and the same is consumed by power plants, industrial boilers, and cement kilns as the fossil fuel supplement. During the tests as conducted by regulatory bodies, the emissions from TDF have not been found to be much different to that obtained from other fossil fuels. The tyres have more energy value per kg than that for coal. It is possible to have the heat value generated equivalent to that obtained from about two gallons of fuel oil by burning a passenger car tyre.
- About 2% rubber in a new tyre is recycled rubber. Hence, it can be considered as the reusable raw materials for tyre manufacture.
- Manufacturing of various molded rubber products e.g. transmission belts, ducts, pipes, automotive parts etc.
- Surfacing of play grounds and stadiums.
- Sound proofing.
- Lining of water storage tanks.
- Other applications of shredded tyres are also possible in civil engineering, crumb rubber producers etc. All these are the beneficial reuse of waste tyres.
- Retreading rubbers consist of about 75% recycled rubber. Further, a truck tyre can be retreaded 203 times on an average before the casing is discarded. Thus, the retreading has proven to be the most effective retardant in scrap tyre generation.

In India, the shredding and crumbing industry needs some kind of legislative support to enforce a wider recycling of the scrapped OTR tyres. Some countries are having a good track record for scrap recycling and have allowed the tyre sellers to charge a nominal fee to cover the cost of transportation of scrap tyres to recycling centers.

f. *Prevention of Scrap Tyres*

Prevention of scrap tyres would be possible by increasing the life of tyres, tyre retreading and newer design of tyres. In Indian context, the following are the recommendations for reducing scrap tyre generation as to counter the threat being posed by the scrap OTR tyres increasing day-by-day.

- A proper upkeep in the use of tyres and an awareness for the right steps in tyre maintenance is crucial to decrease the scrap tyre generation. About 70% of the OTR tyre scrap generation in Indian mining industry is either due to premature unsalvageable cut induced tread separation and blowouts or due to ply separation caused by underinflation. A typical survey of the higher capacity 120T diesel electric driven trucks in India has demonstrated that out of 960 radial tyres, more than 145 tyres went out of service in the range of 0–12000 hours. All such failed tyres have no sign of fatigue in casings except some deep cuts on casings, tread and side walls. The normal overall life of an OTR radial tyre with one or two retreading is reported to range from 9000–18000 hours.

Another factor contributing to ply separation and premature failure of the casing has been due to the exceeding of TMPH rating of tyre.

- Indeed, by promoting the new tyre manufacturing technology and edging out the old type of tyres (Cross ply/Bias) from circulation, the generation of scrap tyres can be reduced. The radial tyres are well known to provide a longer life of tyre casing compared to bias tyres. A longer life of the casing enhances the opportunity for increased tyre retreading.
- Further, timely retreading of tyres can help to reduce the scrap generation because the replaced tread is only 30% of the total weight of tyre.
- Also, by promoting newer tyre design and advanced technology in tyre construction, i.e. the 2-piece tyre design in which the tyre casing is reusable and the tread is replaced whenever worn out. In this design, the bond between reusable tyre casing and replaceable tread is ensured by air inflation. This new technology is expected to cut down the waste generation dramatically because the casing weight is almost 70% of the total tyre weight.

3. Conclusions

- In fact, the tyres are the costly inputs in construction and mining industry. Hence, a proper tyre selection, application and maintenance should be done in a critical manner to meet up the economic viability of mining operation. The recycling of tyres can offset the impact of procurement cost of new tyres to a considerable extent.
- Keeping in view the prevailing scenario in India, 'reduction in scrap generation' seems to be the primary solution not only for the present but for the years to come. However, it should not contradict the resolution proposed by the concerned regulatory government bodies emphasizing various legislations in this context.
- Finally, it should be the endeavour of all concerned employed in the mining industry to promote the recycling of used tyres as much as feasible.

OTR Tyres—Status and Outlook

Jacob George

CEAT Limited, Bhandup
Mumbai-440078

1. Introduction

This paper highlights the following aspects looking from the view point of a tyre manufacturing company:

- Variout OTR applications and its products
- Speciality of OTR operations
- Specific Product requirements and features
- Product development approach
- Observations from the field on tyre life and common failures
- Focus on Customer Service
- Future plans.

The tyre manufacturing company should be fully geared to meet the challenges of the user industries. The tyre manufacturer should have proper technological capability and infrastructural facilities like:

- State-of-the-art tyre material development facility
- High end software for tyre design and development
- State-of-the-art analytical laboratory
- Contact pressure distribution analysis
- Tyre scanning and analysis for
  - reverse engineering
  - re-engineering
- Prototype development and design validation
- Indoor product simulation tests
- Full-fledged indoor product evaluation.

2. Performance of OTR Tyres

In general, the OTR tyres are used in high powered earthmoving vehicles. The performance of such tyres is highly dependant on the kind of application, duty condition, nature of terrain to be negotiated, duty cycle of the earthmover, kind of mining operation etc. It is always recommended to have repair/service facilities near to the site of work for reducing the downtime of earthmovers. Indeed, the downtime of the earthmovers deployed in mining and allied industries is a critical factor.

3. Application of OTR Tyres

In order to design and develop the right type of tyres for use in the industry, it is necessary to have a thorough study on the application of tyres. Further, the design of tyres should be life cycle oriented. The selection of tyres depends on the following major factors.

- Length of haulage
- Speed of vehicle
- Tyre load
- Ground surface.

Off-the-road tyres have various applications in earthmoving machines deployed in mining and allied industries. Fig. 1 presents the possible applications of off-the-road tyres.

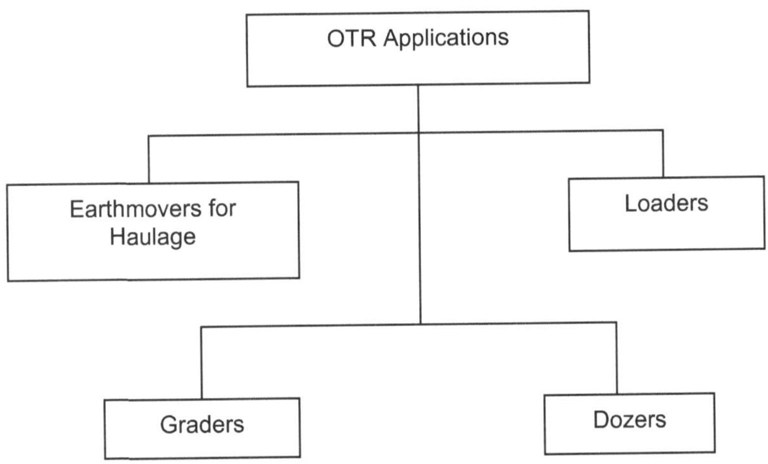

Fig. 1: Possible Applications of Off-the-Road Tyres

The OTR tyres have found wide application in different industries, particularly the mining and allied industries, construction projects etc. These tyres are used in various earthmoving equipment as below:

A. *Earthmovers for Haulage*

The haulage cycle of the earthmoving machines involves primarily one way loading. The capacity of the earthmoving machines has also gone on the increase to handle higher and higher loads commensurate with the increasing production targets of the user industries.

It has become essential for the tyres to have high heat and wear resistance characteristics. The tyres should provide for an improved traction to be suitable for high capacity earthmovers and meet the requirements of covering longer haulage routes.

The Gripmaster ND tyres (Fig. 2) of different sizes as given below would be ideally suitable for use with the earthmovers like dump trucks, articulated trucks, coal haulers, scrapers etc.:

- 14.00-24 GMND E3
- 14.00-25 GMND E3
- 16.00-25 GMND E3
- 18.00-25 GMND E3
- 18.00-25 GMND E4

The special technical features of these tyres are:

- Sturdy lugs
- Highly reinforced casing
- Wide and deep tread
- Unique cap compound
- Special base compound.

The benefits that can be obtained in the use of such tyres are:

- Cut and grinding resistance
- Higher load carrying capacity
- Better traction in all terrains
- Better chipping and chunking resistance
- Cooler running.

Fig. 2: Grip Master ND

The dump trucks are deployed for the haulage of materials on highway or off-highway. These trucks operate within or outside the mining complex. The tyres used on such earthmovers are of the following specification.

- 9.00-20 CT SUPER
- 10.00-20 SUPER DUMPER

Fig. 3: Super and Super Dumper Tyres

Super and Super Dumper tyres (Fig. 3) have special technical features as below:

- Sturdy lugs and buttress
- Improved carcass adhesion
- Extra reinforced tup
- Highly reinforced tread compound.

These tyres are ideally suitable for haulage duty in respect of the following:

- Cooler running
- Improved separation resistance
- Minimum bead failures
- Higher load carrying capacity
- Improved cut resistance and higher mileage.

B. *Dozers*

The dozers are meant to move the materials and are used extensively in mining and construction industries. The main operations of the dozers are pushing and dragging the earth. These earthmoving machines use the Grip Master Loader Dozer tyres (Fig. 4) of size 20.5-25 GMLD L2.

Fig. 4: Grip Master Loader Dozer Tyre

These tyres are characterized by the following technical features:
- Dual angle orientation of lug
- Wide and deep tread
- Highly reinforced casing
- Specially formulated rubber compound for tread.

The tyres can provide the following benefits in the operation of dozers:
- Better directional stability
- Improved traction
- Higher load carrying capacity
- High wear resistance.

C. *Graders*

The graders are widely used for the construction and maintenance of haul roads. It is necessary to meet the following requirements in the selection of tyres for such earthmoving equipment.
- High traction
- Improved cut resistance
- Better directional stability.

The Grip Master Grader tyres (Fig. 5) available in following sizes for possible use in grading vehicles:
- 13.00-24 GMG G2
- 14.00-24 GMG G2
- 15.5-25 GMLG G2
- 17.5-25 GMLG G2

Fig. 5: Grip Master Grader Tyre

The special features of these tyres are:
- Orientation of lug
- Wide and deep tread
- Specially formulated tread compound.

The tyres can provide the following benefits in the operation of graders:

- Better directional stability
- Improved traction in all terrains
- Improved wear and cut resistance.

D. *Loaders*

The loaders have found a wide application for the purpose of loading and transfer of earth and rocks over short distance. Generally, these machines operate at a low speed and can negotiate rough terrain effectively. The Grip Master loader tyres (Fig. 6) are available in different sizes as below:

- 15.5-25 GMLG L2
- 17.5-25 GMLG L2
- 20.5-25 GMLD L2
- 18.00-25 GMND L3

Fig. 6: Grip Master Loader Tyre

The technical features of these tyres are:

- Dual angle lug orientation
- Wide and deep tread
- Highly reinforced casing
- Superior quality of tread compound.

The benefits expected in the use of such tyres are:

- Better directional stability
- Improved traction
- Higher load carrying capacity
- Improved cut and wear resistance
- Cooler running.

E. *Underground Mining Loaders*

The load-haul-dump machines are popularly used in underground mines for loading, haulage and dumping of minerals or other solid wastes.

The tyres used in such machines are of type 12.00-24 SLEEK (Fig. 7). These tyres are provided with highly reinforced casing. The tyres have high load carrying capacity and can resist cut and wear effectively.

Fig. 7: Sleek Tyres

4. Field Observations

The tyres in use would show a wide variation in life. Depending on the conditions prevailing at the site of work, there are incidences of cuts, chipping etc. Generally, the types of tyre failures observed in the operation of earthmoving equipment are:

- Through cut
- Side damage
- Impact break
- Damage due to improper fitment.

When the tyre temperature goes beyond the recommended values during the running of vehicles, it may lead to tyre failure.

5. Status and Outlook

The manufacturer have been striving hard to bring in continuous improvement in tyre design, manufacturing and materials of composition. The tyre manufacturers have been using successfully the statistical quality control techniques accompanied by product and process rating systems. Further, the tyre making companies has on the agenda to establish the concept of quality circles. Consequent to this, the tyre manufacturer e.g. CEAT has been successful to receive the ISO-9001 accreditation in 1995 followed by ISO/TS16949 : 2002 as the forerunner company in India.

The tyre manufacturers are geared to provide the right type of customer oriented solutions and maintain a regular interaction with the manufacturers of earthmoving machines. Much attention is being paid by the tyre manufacturers to develop an effective after sales service system. Consequent to this, the tyre companies are offering a rigorous on site training to the operating personnel of organizations using the tyres. Apart from this, an endeavour is always made to provide faster and prompt response to the users as a measure of after sales service.

The R&D departments in tyre manufacturing industries are engaged in the design and development of newer materials, reinforcements and rubber compounds for the manufacturing of most suitable tyres for deployment at the user industries.

The aims of the tyre manufacturers are to implement the computer aided engineering system for an improved product development and subsequently upgrade the product testing facilities commensurate with the adoption of newer technology. The objective of the tyre manufacturers is to meet the increasing demand of the off-highway-tyres on one hand as well as to extend a high value service to the customers on the other and finally pave the way for a successful participation in the global markets.

Application of OTR Tyres

R. Sengupta
JK Vikrant Tyres
7, Council House Street, Kolkata-700001

1. Introduction

This paper aims to discuss on some basic aspects related to the applications of OTR tyres, factors affecting the tyre life in operation and maintenance etc.

Off-the-Road (OTR) tyres are generally large and heavy tyres used in various earthmoving equipments. The OTR tyres can also be used on various other special purpose vehicles like Mobile Cranes, Container handlers, Airport emergency vehicles etc. Because of the severity of working environment, the design and manufacturing of OTR tyres generally utilize the highest level of technology available in tyre manufacturing industry.

2. Application of OTR Tyres

A. *Graders*

The primary function of the graders is to level the surface. The graders are usually provided with rear wheel drive. Also some of the graders can articulate. The size of tyre used in dozers is 13.00–24 12PR G2.

B. *Loaders*

The main function of the loaders is to load bulk materials into other vehicles. The loaders are most commonly used amongst the earthmoving vehicles. The typical tyre sizes used in loaders are:

14.00–25 20PR E3
26.5–29 28PR L3
29.5–29 28PR L3

C. *Scrapers*

This haulage vehicle is often used in construction and reclamation work to provide an even distribution of load over a given area. The typical tyre sizes used in this machine are:

23.5–25 E3
18.00–33 E4

D. *Rigid Frame Haul Trucks*

The primary function of these trucks is to haul materials from one point to another. The payloads of these vehicles may range from 25 to 300 tons and may go even higher. These vehicles are diesel powered with either electric or mechanical drive. The typical tyre sizes used in these machines are:

18.00–25 32 PR E3/E4

18.00–33 32 PR E4

21.00–35 36/40 PR E4

24.00–49 48 PR E4

36.00–51 52 PR E4

E. *Articulated Haul Trucks*

The articulated trucks are considered as high mobility haulage vehicles. These vehicles are provided with all wheel drive. The design features of the machine enable to negotiate a short turning radius. The payload of these vehicles generally varies in the range of 20–50 T. The typical tyre sizes used in these machines are:

18.00–33 32 PR E4

20.5 R 25

3. Industry Codes

The industry codes in the use of tyres are given below for different earthmoving vehicles:

a. Earthmover

E1–Standard tread rib

E2–Standard tread traction

E3–Standard tread rock

E4–Deep tread rock

E7–Standard tread flotation

b. Dozer and Loader

L2–Standard tread traction

L3–Standard tread rock

L4–Deep tread rock

L4S–Deep tread smooth

L5–Extra deep tread rock

L5S–Extra deep tread smooth

c. Grader

G1–Standard tread rib

G2–Standard tread traction

G3–Standard tread rock

G4–Deep tread rock

4. Tyre Nomenclature

18.00–25 32 PR VEM 99 E4 HRC/CRC 2S

18.00	:	Nominal section width of tyre inch
–	:	Bias type construction replaced by 'R' in radials
25	:	Nominal rim diameter, inch
PR	:	Ply rating
VEM 99	:	Vikrant Earth Mover 99 Family
E4	:	Industry Code
HRC	:	Heat resistant compound
CRC	:	Abrasion cut resistant compound
2S	:	HRC ← (–) S (+) → CRC

5. Tyre Selection

The basic factors to be considered in the selection of tyres are :

a. Physical size : The tyre must fit within the "Wheel Well".

b. Carry the load : The tyre must be able to handle the specified payload of the vehicle commensurate with the Gross Vehicle Weight (GVW).

c. Minimum traction : The tyre must be able to meet the minimum traction requirements.

The other parameters related to vehicle operation should also be considered as given below:

- Length of haul
- Cycle time and speed of vehicle
- Work schedule
- Grades to be covered
- Curves to be negotiated
- Underfoot conditions
- Actual load and load distribution.

6. Inflation of Tyre

Maintaining the proper inflation pressure is the most important requirement in a tyre maintenance program. The published inflation pressures are meant for standard loads. When marginal overloads are to be encountered, the inflation pressure should be increased by 2% for each 1% of the overload.

The maintenance of tyre inflation pressure should be done in respect of the following:
- Overinflation
- Underinflation
- Proper inflation

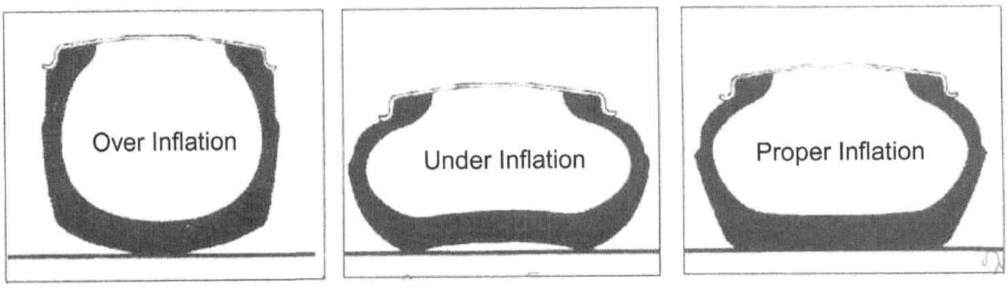

Fig. 1: Tyres in Various Stages of Inflation

The overinflation subjects the center of tread to excessive wear, whereas the underinflated tyres flex excessively at every turn of the wheel resulting in high internal heat and premature failures. Maintenance of proper inflation of tyres goes with the correctly inflated tyres that permit all portions of the tread to be in contact with road, yet the tyres are not soft enough to flex excessively.

Maintenance of proper tyre inflation goes with the correctly inflated tyres that permit all portions of the tread to be in contact with road, yet the tyres are not soft enough to flex excessively.

6.1 *Tyre Inflation Programme*

A sound tyre inflation programme is governed by the following:
- To train personnel for maintaining the proper tyre inflation.
- To check that cap is provided on every valve
- To ensure that accurate pressure gauges are used
- To perform cold checking of tyre whenever possible, may be weekly in ideal condition
- To ensure daily checking of hot pressure.

Apart from the above, the maintenance of tyres and its record keeping are also the important factors to be considered.

7. Dual Tyre Assembly

The following factors should be considered in the assembly of dual tyres:
- Tyres of same diameter/size
- Tyres of same construction
- Tyres of same manufacturer
- Tyre of same construction across an axle.

8. Tyre Life

The life of tyres depends on the soil condition of the haul road, grades along the haul road and the curvatures to be negotiated as detailed in Table 1, 2 & 3 in respectively.

Table 1: Tyre Life and Haul Road Soil Condition

Soil Condition	*Tyre Life (%)*
Soft rock—no rock	100
Soft rock—some rock	90
Blasted sharp rock	60
Water and sharp rock	40

Table 2: Tyre Life and Haul Road Grades

Grade (%)	*Tyre Life (%)*
0	100
6	90
10	80
14	65
18	50

Table 3: Tyre Life and Haul Road Curves

Curves	*Tyre Life (%)*
None	100
Medium	90
Sharp/severe	80

Use of OTR Tyres in Bharat Coking Coal Limited

V.B. Singh

Bharat Coking Coal Limited
Dhanbad-826005

1. Introduction

Coal India Limited is one of the largest consumers of OTR tyres in the country. On an average, Coal India Ltd. has been procuring annually the OTR tyres of different sizes of worth Rs. 40.00 cores. The specification of tyres range from 11.00 × 20.00 to 36.00 × 51.00. Bharat Coking Coal Limited (BCCL), a subsidiary of Coal India Limited, uses tyres of various sizes varying from 11.00 × 20.00 to 24.00 × 49.00 of value Rs. 5-6 crores. A large number of tyre mounted earthmoving machineries have been deployed at the opencast mines of BCCL.

2. Deployment of Dumpers

About 38 nos. of 85T dumpers have been deployed at the Block II mines. In fact, the dump trucks deployed at this mine are all of 85T capacity. Also, the 85T dump trucks have been operating at the Jamunia Opencast mine of Barora Area. In Jamunia mine, 10 nos. of 85T dumpers and 15 nos. of 35T dumpers are in operation; whereas the Muraidih mine has been running with 12 nos. of 85T dumpers and 20 nos. of 35T dump trucks.

Most of the opencast mines of BCCL are in operation using the dump trucks of 35T capacity. The relatively larger fleet sizes of 35T dumpers having 30-36 (approx.) trucks in each fleet are deployed at the major opencast mines of BCCL (Table 1).

Table 1

Sl. No.	*Name of Area*	*Name of Mine*
1.	Govindpur	Block IV
2.	Katras	West Mudidih/Keshalpur
3.	Sijua	Nichitpur
4.	Bastacolla	Ghanoodih
5.	Chanch Victoria	Dahibari/New Laikdih
6.	Kusunda	Kusunda

Apart from the mines mentioned above, some opencast mines of BCCL (Table 2) are in operation with a moderate fleet size of 35 T dump trucks i.e. about 20–28 dumpers in each fleet.

Table 2

Sl. No.	*Name of Area*	*Name of Mine*
1.	Kustore	South Jharia (Rajapur)
2.	Barora	Muraidih
3.	Katras	New Keshalpur
4.	Bastacolla	Golakdih
5.	East Jharia	Damodar
6.	Lodna	North Tisra
7.	Lodna	South Tisra

Further, some other surface mines of BCCL (Table 3) have been running with 35T dump trucks with comparatively lesser fleet sizes i.e. 10–18 dumpers in each fleet.

Table 3

Sl. No.	*Name of Area*	*Name of Mine*
1.	Block II	Jamunia
2.	Barora	Damoda
3.	Sijua	Sendra Bansjora
4.	Sijua	Tetulmari
5.	Kustore	Ena
6.	Bastacolla	Golakdih
7.	Bastacolla	Bastacolla
8.	Lodna	Jeenagora

Moreover, some newer opencast mining projects have been opened in BCCL (Table 4), while the fleets of 35T dump trucks operating at such projects comprise of a few vehicles only i.e. 4–6 dumpers in each fleet.

Table 4

Sl. No.	*Name of Area*	*Name of Mine*
1.	Kusunda	Khas Kusunda/Gondudih
2.	Kusunda	Dhansar
3.	Govindpur	Jogidih
4.	Barora	Satabdih

Generally, all the surface mines of BCCL are provided with 1–2 Water Tankers for the purpose of dust suppression at the haul roads and other adjoining areas. Depending on the capacity of dump trucks used in a particular surface mine, the water tankers of equivalent sizes are deployed. Indeed, the mobile water tankers operating at different opencast mines are the modified versions of 35T or 85T dump trucks so that tyres of same size can be used to streamline the tyre inventory management as well as maintenance replacement activities.

Table 5 shows the opencast mines in BCCL, where moderate-to-large numbers of dump trucks have been deployed for coal haulage and disposal of overburden.

Table 5

Sl. No.	*Name of Area*	*Name of Mine*	*No. of Dumpers*	*Dumpers Capacity*
1.	Block–II	Block–II	38	85T
2.	Govindpur	Block–IV	36	35T
3.	Chanch Victoria	Dahibari/New Laikdih	36	35T
4.	Barora	Muraidih	12	85T
			20	35T
5.	Bastacolla	Ghanoodih	32	35T
6.	Sijua	Nichitpur	30	35T
7.	Katras	West Mudidih/Keshalpur	30	35T
8.	Kusunda	Kusunda	30	35T
9.	Lodna	South Tisra	28	35T
10.	Katras	New Keshalpur	26	35T
11.	Block–II	Jamunia	10	85T
			15	35T
12.	East Jharia	Damodar	24	35T
13.	Kustore	South Jharia (Rajapur)	22	35T
14.	Lodna	North Tisra	22	35T

Evidently, a huge number of OTR tyres are involved in the operation of such large number of fleets of 35T and 85T dumpers in different surface mines of BCCL.

3. Tyre Life

The expenditure on the procurement of tyres in this company is a significant factor and this comes as the major expenditure on a particular consumable item. Hence, the life of the OTR tyres while in service plays an important role in cutting down the cost of production. This paper has aimed to highlight the aspect of average life of new tyres before retreading as well as that after retreading.

It has been observed in BCCL that the average life of a new tyre before doing any retreading operation is about 4000 hours. The average life after first retreading is 2000 hours and that after 2nd retreading is 1500 hours. Normally, it is seen that every year about 60% of tyres are found fit to undergo retreading after the completion of 1st useful life, whereas about 20% of these retreaded tyres are found fit for carrying out 2nd retreading. The balance requirement of tyres is being met through the procurement of new tyres.

4. Tyre Maintenance and Care

Although the maintenance of tubeless tyres is easier as compared to tubed tyres, it has to be remembered that the tubeless tyre has to be discarded without any retreading once it gets a through-cut. On the other hand, the tyre manufacturers claim that the OTR tyres are having the salient features as below during the period of its service life:

(a) Tread reinforcing bar
(b) Through cut resistant tread compound
(c) Heavy sidewall construction
(d) Heavy buttressed lugs
(e) Heat resistant layer under tread
(f) Rugged nylon construction to achieve the optimum tyre life.

Despite the claims made by the manufacturers with regard to above technical features of OTR tyres, a large percentage of new tyres are still failing while in use at the coal mining complexes before achieving the optimum life. As the above issue is of great concern to this coal mining subsidiary, it is prudent to place this issue at a forum where both the tyre manufacturers and the users are participating so that a meaningful solution can emerge in terms of mutual benefit.

M/s TIP TOP General Agency, New Delhi has established the technology for repairing the through cut tyres and such repairs made by this company were found to be satisfactory. Thus, BCCL is able to reduce the expenditure to some extent by resorting to such repair of tyres especially for the larger sizes of tyres.

5. Conclusions

As the user of tyres, the mining companies should take proper care to enhance the tyre life. The general factors involved in this connection are:

(a) Proper storage of tyres

(b) Proper inflation of tyres

(c) Maintenance of haul roads to avoid through cuts and spinning of tyres

(d) Dressing of tyres to avoid further damages due to external injury.

Finally, the disposal of scrap tyres poses to be the other problem. As on date, the scrap tyres are being disposed off by auction through MMTC. However, this is a complex process and it takes considerable time. In case a system can be developed that the manufacturers take back the scrap tyres by fixing a proper salvage value for different sizes of tyres, the system would be easier as well as quicker with regard to disposal of scrap tyres.

Attainment of Agility through Just-in-Time Approach in a Premier Truck-Tyre Manufacturing Unit

S. Chattopadhyay, A.K. Basu, Indira Prasanna and Reena Bajaj

Indian School of Mines
Dhanbad-826004

1. Introduction

Today the industries have been facing tremendous competition in their functioning. In order to maintain a competitive edge, the manufacturing industries are faced with the task of reducing costs and improving the quality of products. Consequently, it has become necessary to make the manufacturing processes effective, fast and burdened with low overhead and workover jobs.

The conventional design processes appear to be rigid when the product features undergo frequent change. In the present era, it is necessary that the quality and reliability of the product is assured and the time lapse between the placement of order and complying the delivery should be minimized. In brief, any change in product design should be incorporated at a rate even faster than the market competition.

The modern industries have been witnessing the revolution in different areas starting from design to manufacturing. Agile manufacturing is one of the most modern concept in manufacturing, where the enterprise learns to be proactive and to work in a continuous manner. It is the total systems approach that develops a culture in which everyone is engaged in improving the production processes and the production target thereof continuously. It leads to integration of vision, lower cost attainment and a shorter delivery time. Agile manufacturing will enable the industries to offer better service, quality design and flexibility to compete in the real world market. It helps in manufacturing the high quality products efficiently and subsequently provides flexibility to the companies to accommodate the changes in design, product and product mix quickly and efficiently without affecting quality and cost. This paper describes the attainment of Just-In-Time practices in a premier truck tyre manufacturing unit.

Just-In-Time (JIT) is one of the important methodologies adopted for the control of modern inventory system. This concept has emerged to fulfill the demand of user-market-environment. It may be viewed as the production system, aiming to improve the overall productivity through elimination of waste and leading to an improved quality of the product. More precisely, JIT is a simple and efficient technique that goes with a minimum of wastage. In an ideal JIT system, the inventory at each stage for a perfectly balanced line should be zero; while the maximum inventory in any system should not be more than one unit. JIT is practiced through the identification of key process areas by studying critically the variability in the product-mix.

In this paper, a detailed study has been made on the existing system keeping in view the drawbacks of truck tyre manufacturing unit. Finally, the merits of the JIT practices are highlighted through the support of necessary statistical data.

2. Truck Tyre Manufacturing System

In this study a premier industrial truck tyre manufacturing industry has been selected for case study development for the implementation of JIT in the light of attaining the agility in existing system. The characteristics/features of the manufacturing system are presented below (Fig. 1):

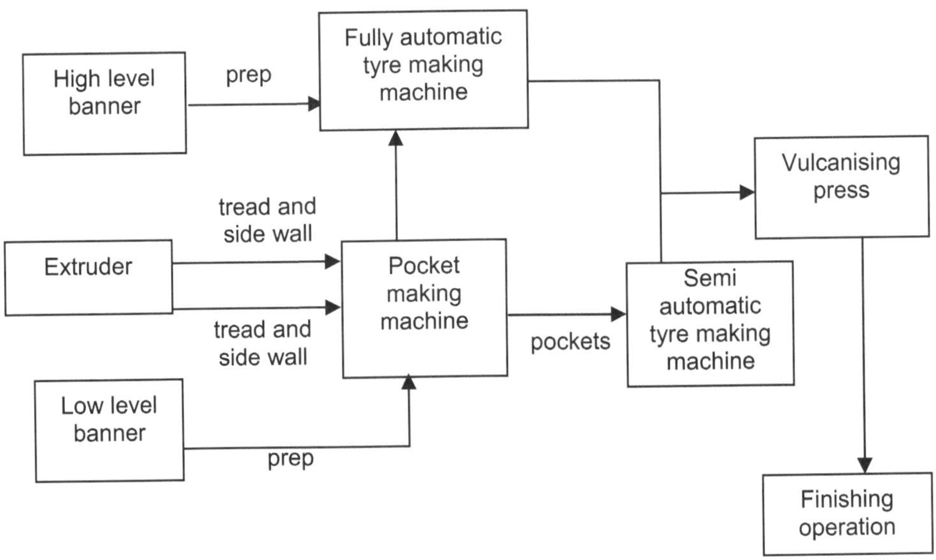

Fig. 1: Layout of Truck Tyre Manufacturing System

a. High Level Banner

The high level banner is used for bias fabric cutting in automatic tyre making machines. Apart from the truck tyres, the fabric for the aero tyres and Off-the-Road (OTR) tyres are also cut in the form of tube by this unit. An optical sensor is provided to identify the presence of rubberized fabric and actuate the cutter accordingly.

b. Low Level Banner

This unit is primarily utilized to cut fabric for the pockets of truck tyres. The truck tyres are manufactured by means of the semi-automatic and the general-purpose tyre making machines. This unit does not have to feed the automatic tyre making processes using the rubberized fabric.

c. Extruder

This machine extrudes the tread rubber alongwith the sidewall rubber in one run. This is a special type of extruder through which two different kinds of rubber can be extruded in one integrated layer for making the truck tyre. This unit is required as the constituents namely tread rubber and sidewall rubber is of different materials to be used in a truck tyre.

d. Tyre Making Machine

In tyre manufacturing factory generally three types of tyre making machines are used as given below:

i) *Semi Automatic Tyre Making Machine*

This machine requires specified pockets for different forms of tyre construction. It has a segmental collapsing device for removal of the raw tyre from the machine. As compared to the Zero collapsing machines, the only problem in the use of this machine is that all the segments have to be unloaded alongwith the raw tyre. Thereafter, the segments have to be mounted again on the machine for making the next tyre.

ii) *Zero Collapsing Machine*

These are some special kind of tyre making machines that enable an easy removal of the previously made raw truck tyre. The productivity of these machines is higher than the general purpose segmental type machines. However, like the general purpose tyre making machines, these machines also need pockets of rubberized fabric for manufacturing raw tyre. The benefit in using these machines is that their unloading time is lesser to a great extent in comparison to the earlier units.

iii) *Automatic Tyre Making Machine*

These are special purpose tyre making machines for making raw tyres without needing the pockets as required for the other types of tyre making machines. The productivity of these machines is much higher than that of the other types of machines. This is due to the reason that these machines do not require pocket-building operations in the manufacturing of raw tyres. Hence, the complexity in manufacturing, requirement of setup time and manpower required in the process get reduced substantially.

e. Vulcanising Press

These presses are responsible for vulcanizing the raw truck tyre with circulating hot water and steam flow during vulcanization. The profiles of the moulds provide the desired shape of the truck tyres. The moulds are made of aluminum. The flow rates of water and steam are controlled through solenoid valves for a pre-determined period of time.

It is revealed from the Fig. 1 that the fully automatic tyre making machine does not require pockets of rubberized fabric; while the other group of machines provided for tyre making requires pockets of prep as input in the manufacturing of raw tyre. These raw tyres come from both the types of tyre making machines and are cured in the vulcanizing press. After the cured tyre undergoes finishing operation, these are dispatched to the stock area of finished goods.

3. Diagnostic Review of Tyre Manufacturing System

i) High Manufacturing Cost

Due to various reasons the manufacturing cost of the tyre making unit is high. One of the reasons is that a good number of skilled workforce is required in tyre making and mould preparation operation. The overhead expenditure is high as the layout is not planned properly and the support services are also provided in an unplanned manner. As a result of the high setup time and the requirement of a lengthy procedure for the incorporation of design changes, the overhead expenditure goes on the higher side significantly.

ii) Lack of Traceability

Due to the stockpiling of goods, maintenance of the Stock-in-Progress (SIP) becomes difficult and cumbersome. Thus, to make a trace of goods it becomes necessary to employ an extra workforce resulting in higher investment. Also, some loss is incurred due to the damage of piled-up goods.

iii) Blockage of Money

Due to longer die changing time, the marketing of products takes comparatively longer time in such manufacturing unit. This results in the retardation of cash flow and thereby reduces the profit margin.

iv) Pilferage of Goods

A large inventory of stock leads to undesirable practices like pilfering and smuggling. Consequently, the growth of the system as a whole is hampered.

v) Lack of Agile Response to Market Demand

Due to the accumulation of stock vis-à-vis high level of inventory, the attainment of agility is not possible as the procedure becomes uneconomical. The manufacturer cannot respond quickly to the change required in the existing system. It becomes difficult to accommodate the changes desirable in lotsize and production quality.

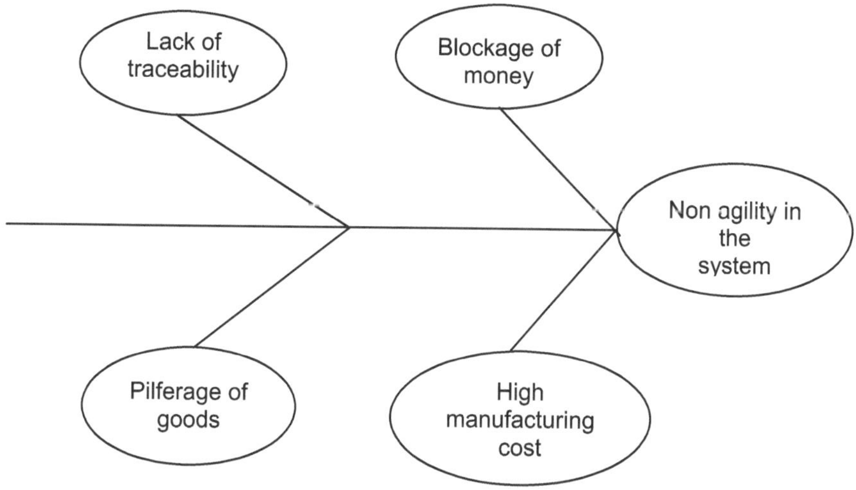

Fig. 2: Fish-bone Diagram Representing System Non-Agility

A fish-bone diagram as in Figure 2 would represent the problems of the truck tyre manufacturing unit.

4. Results and Discussions

It is felt that there is enough scope for applying the concept of JIT and it should be exploited for an effective management of the working capital needed in tyre manufacturing organizations. Today, manufacturing organizations are facing a high level of competition and are striving for growth and making profit with a minimum of product life cycle. One of the most important methodologies adopted for modern inventory control system is JIT. It not only enables a production unit to operate at a minimum possible inventory level; but also brings substantial ease in the operation of material handling devices with optimal batch load. In fact, JIT is a revolutionary concept due to its simplicity for an effective management of the working capital. It introduces no advanced technology or any complicated principle; rather strives to eliminate the unnecessary burden of complexity in the production system leading to the development of a minimum inventory holding. No complicated process is necessary in JIT. An industry can build the same product with lesser labour, lesser overhead with improved quality and in lesser time by using the principle of JIT system to outsmart the competitors. Making the complex situation simple is the main goal of JIT. The positive results that can be obtained in the area of working capital management are as below:

- The entire manufacturing system can be made operational with nearly one-third inventory level in comparison to the traditional system;

- The traceability of materials at any corner of the layout becomes considerably easier;
- The manufacturing system operates with lesser load on the material handling equipment and results in minimum handling cost;
- The scheduling of material handling facilities becomes considerably easier in comparison to the traditionally practiced processes;
- Lesser traffic load on the paths and routes of battery operated power trucks results in reduced operational time and cost;
- Wastage is minimized due to the ease of supervision of the work-in-progress stock of smaller quantity;
- The development of any wastage due to sudden changes in priorities is reduced resulting in an aggressive working capital policy by maintaining the minimal inventory safely.

5. Goal Performance Monitoring

JIT can be followed gainfully in an agile manufacturing system. With the application of this methodology all kinds of stocks and storages are reduced to a minimum level due to the implementation of short-cycled manufacturing operations. With the application of JIT practices for an extruder unit, the number of treads manufactured per shift can be increased substantially.

The Tables no. 1 & 2 present a comparative study on the number of treads per shift in truck tyre manufacturing unit using various tyre making machines with one and two extruders respectively.

Table 1: With One Extruder

Sl. No.	*Tyre Making Machine*	*No. of Treads per Shift*		
		Low	*Medium*	*High*
1.	Automatic	5	20	50
2.	Segmental	20	50	80
3.	Zero Collapsing	20	50	80
4.	Traditional	15	40	60
5.	Small OTR	2	5	10

Table 2: With Two Extruders

Sl. No.	*Operation*	*Die Changing Time (min) for Making Tread*		
		Low	*Medium*	*High*
1.	Bolt/Nut opening	1.5	0.5	0.1
2.	Die unloading	4	2	0.3
3.	Die loading	4	2	0.3
4.	Bolt/Nut hightening	1.5	0.5	0.1
	Total time taken	11 min	5 min	1 min

The graphical representation of the data as Table 1 has been shown in Fig. 3.

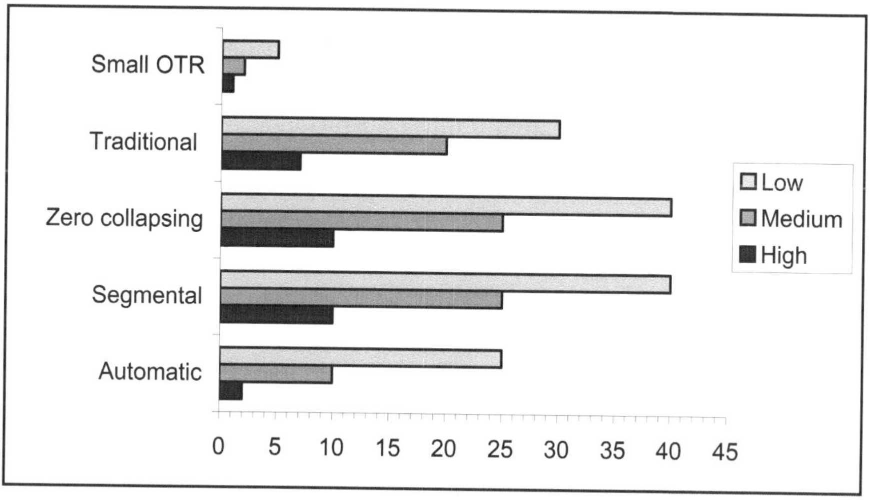

Fig. 3: No. of Treads per shift

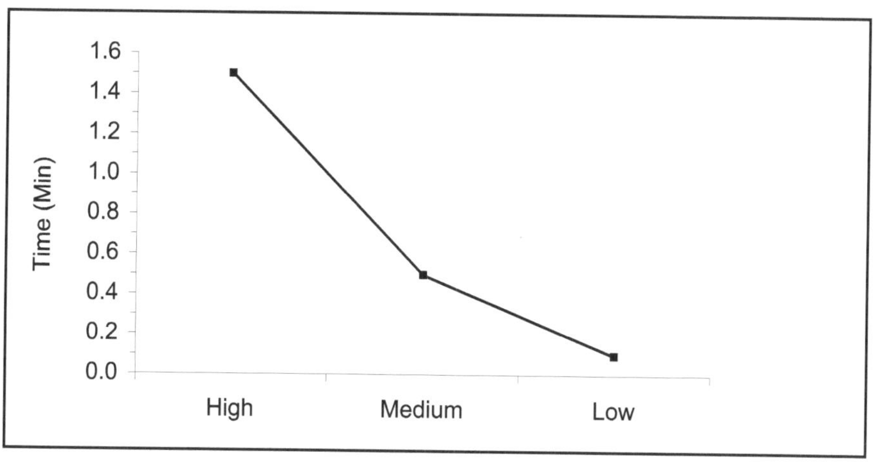

Fig. 4: Opening Nuts and Bolts

An analysis of the above tables and graphs reveals that the no. of treads per shift for each type of tyre making machine would increase by 4–5 times from the low to high values of no. of treads/shift and that by 1.5–2 times from medium to high values of no. of treads/shift.

In this study of truck tyre manufacturing unit, it is observed that for a minimum inventory in operation, the implementation of JIT process would be much useful. It is essential to concentrate on the scope for reducing the die changing time to attain the agility in tyre manufacturing system. The following components of the die changing operation are to be considered for having the idea of work measurement:

1. Opening the nuts and bolts (Fig. 4),
2. Unloading the existing die(Fig. 5),
3. Loading the new die (Fig. 6),
4. Tightening the bolts nuts (Fig. 7).

Table 2 shows the die changing times for different operations to perform the task for low, medium and high number of treads/shift.

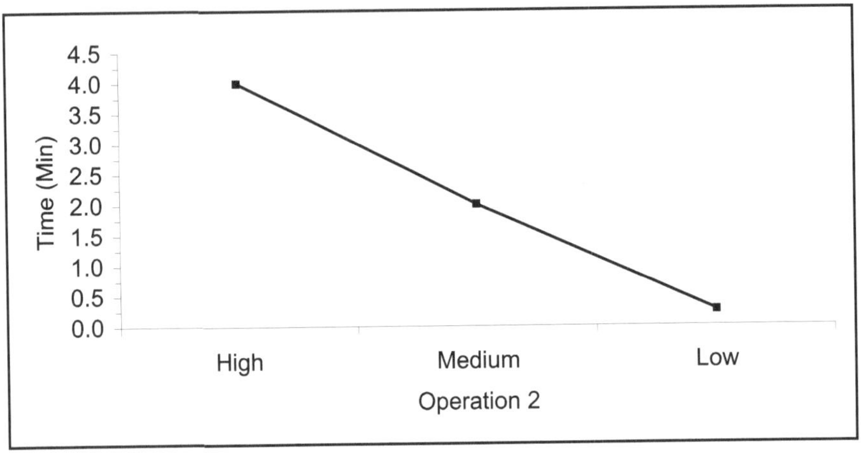

Fig. 5: Unloading Existing Die

Figures 8 and 9 show a comparative study on the die changing time for various operations using one or two extruders.

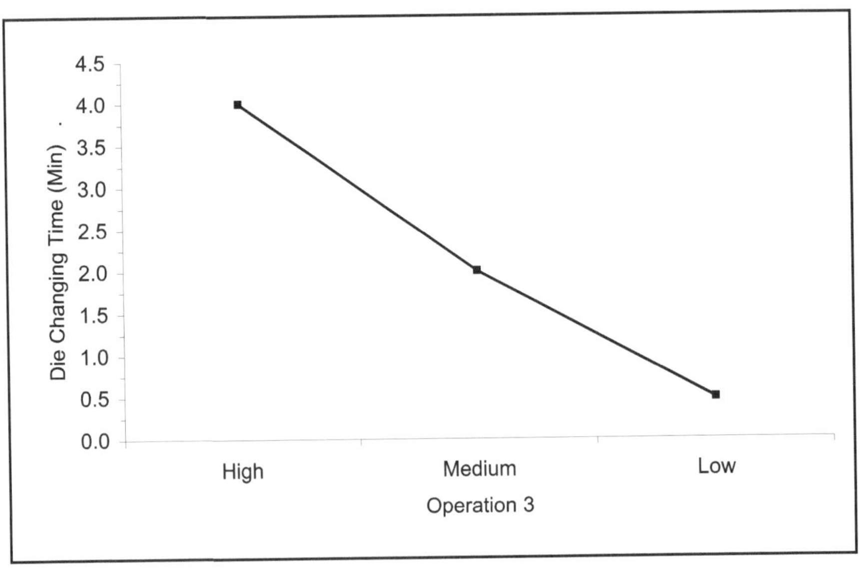

Fig. 6: Loading New Die

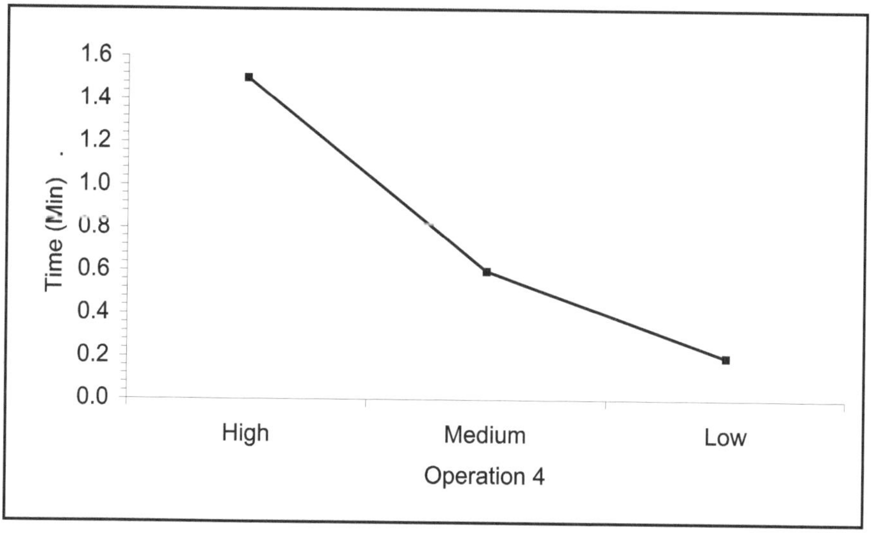

Fig. 7: Tightening Bolts and Nuts

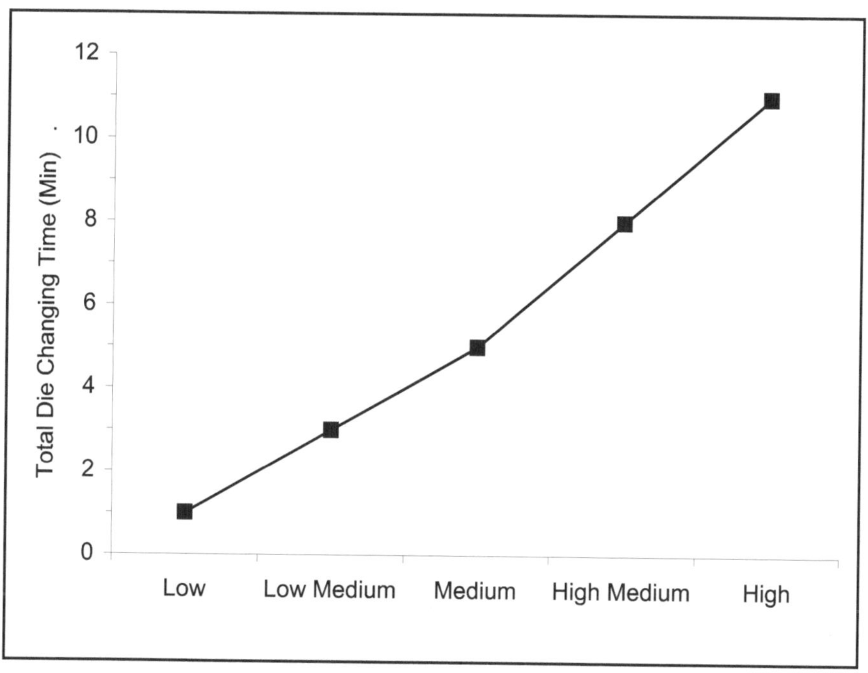

Fig. 8: Total Die Changing Time with One Extruder

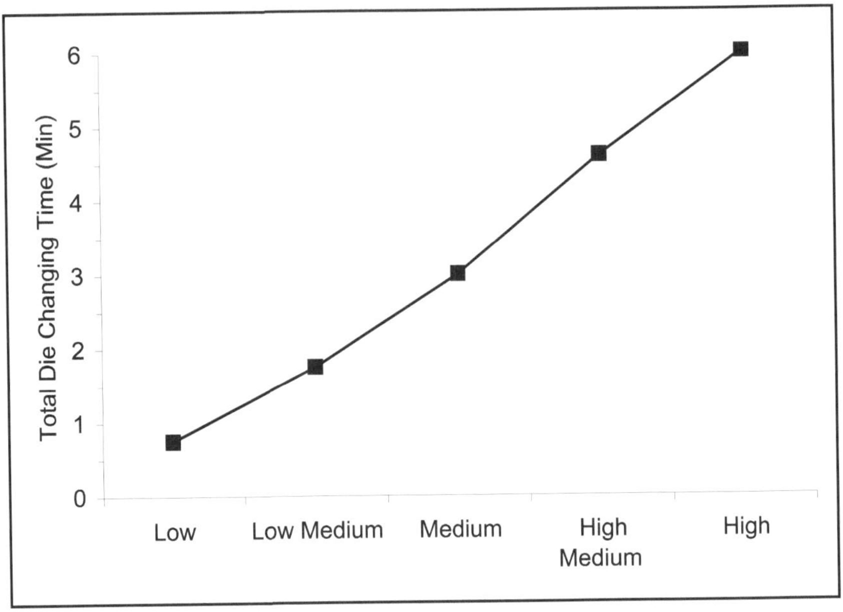

Fig. 9: Total Die Changing Time with Two Extruders

However, it has not yet been possible to achieve the zero inventory in real life. The matter is still in the development stage. Anyway, the manufacturing industries could aim to achieve Single Minute Exchange of Die (SMED). It is one of the techniques that would support the short cycle characteristics of the technology. The possible methodology may involve:

- Use of power spanners,
- Incorporation of slider die magazines,
- Fitment with taper attachment of dies,
- Gang tree based dedicated loading and unloading of die magazines.

In a truck tyre manufacturing unit, the product variety is substantially large. Hence, the application of one extruder may not provide considerable operational efficiency accompanied by various complexities. The sensitivity analysis for the reduction in die changing time is crucial with regard to the augmentation of the manufacturing facilities. The reduction in die changing time would enable the manufacturing unit to become a candidate for the implementation of JIT and for running the plant at the minimum possible cost with least hazards.

A Novel PC-Based Cure Monitoring Process of OTR Tyres

P.K. Sadhu[1], S. Chattopadhyay[2] and D.K. Mittra[3]

[1, 3]Department of Electrical Engineering
[2]Department of Mechanical Engineering & Mining Machinery Engineering
Indian School of Mines, Dhanbad-826004

1. Introduction

In the real life case study, the complicated moulds of the bag-o-matic curing (vulcanizing) presses are considered to be manufactured through intelligent built-in-diagnostic based flexibly printed micro-circuit layer network of distributed piezo-electric actuators [1, 2]. These actuators will establish sensitive micro-cure monitoring and intelligent on-line knowledge based corrective vulcanization procedures of raw rubber products. The aim of this paper is to suggest built-in-diagnostic [2, 3] features for implementation of intelligent cure monitoring process of rubber products in the micro manufacturing environment. Proposed experimental set-up will be useful for improving the quality of curing of rubber related products in the vulcanizing presses. In order to build smart [3, 4] and intelligent systems materials must posses some extraordinary properties in comparison to the traditional materials. In actual condition, mechanical and thermal stresses have to be sensed. In case of the smart materials the degrees of the intensity of the environment parameters have to be sensed. On-line [7, 8] corrective measures also have to be taken through actuators.

There are many types of sensors [5, 6] that can detect failure in the composite material but implementing a large network of sensors on new or existing structures poses a big challenge for developing a practical monitoring system. Hence, the network of sensors must be easily and reliably integrated with the host structure so that there is no effect on the integrity of the structures. Methods using the principle of flexible circuit printing technique have been developed which implement a network of distributed piezo-electric sensors. This smart flexible layer is made of a die electric cell with a distributed network of piezo-electric disc serving as sensors [2–6]. The main design criteria are:

- The layer must be able to withstand typical composite cure temperature.
- The layer must provide adequate electric insulation for the embedded wires and devices.
- The layer must have minimal effects on the integrity of the host structures.

A case study has been taken for premier Indian rubber manufacturing unit that has two major divisions. One is tyre and the other is the industrial product division. The tyre division is emphasized for the work as this division provides the bulk of the turn over and the profit for the companies. After analyzing a set of questionnaire is prepared for determining the agility of that organization. The following demerits of the company have been identified on the basis of previous analysis.

(i) Comparatively high rejection cost.
(ii) Inconsistency in product quality.
(iii) Comparatively non-proactive machineries.
(iv) Comparatively longer inspection time.
(v) Pressure cure monitoring is minimum.
(vi) Temperature cure monitoring is not adequate.
(vii) Actuators are guided manually.
(viii) The defects like bag mark, over curing zone, under curing zone, shape deformation are very common.

From the above analysis it has been observed that most of the problems of this premier tyre manufacturing company can be reduced to a substantial amount through a novel PC-based cure monitoring process.

2. Proposed Scheme

A PC based cure monitoring system suggested in the paper comprises of:

(i) Smart layer based sensing.
(ii) Solenoid valve based actuation.
(iii) Piezo-electric sensor based pressure measurement in the mould.
(iv) Reliable on-line data acquisition.
(v) PC interfacing.
(vi) Software development for PC based control.
(vii) Control room based monitoring.

Figure 1 shows the digital data acquisition system. Pressure, temperature and flow sensors/transducers convert physical quantity into electrical signals, which are acceptable by the data acquisition system. Signal conditioners are used for processing the transduced low amplitude signals into a usable format for the measurement. Sequential channel selection is done by means of a 16-channel analogue multiplexer. The multiplexer can handle sixteen single ended analogue signals. The switching-time of the multiplexer is software controllable and can be set/adjusted for optimum performance. The output of the multiplexer which represents one transducer signal at a time is suitably amplified by an instrumentation amplifier. The gain of the amplifier is also software adjustable in steps. A 12-bit Analogue to Digital Converter (ADC) is used to convert the analogue signals into the 12-bit digital format with good resolution.

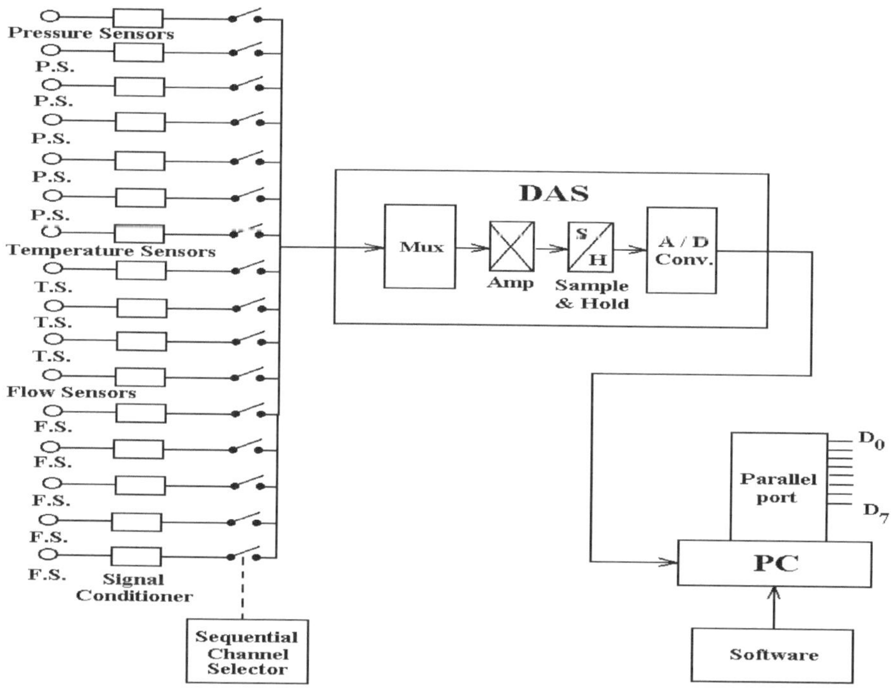

Fig. 1: Data Acquisition System for PC-based Cure Monitoring for OTR Tyres

The digital output of the ADC is down loaded to the PC. A Dedicated Data Acquisition (DAS) card, which is a combination of analogue multiplexer, sample-hold, amplifier and analogue to digital converter resides inside the PC and uses the internal power supply of the PC for its operation. It is well known that the signal fed to an Analogue to Digital Converter (ADC) should be maintained constant during the conversion period. The sample and hold circuit takes care of this and provides a constant input to the ADC during conversion period. For 16 inputs to the 16-channel analogue multiplexer, any one of the inputs can be obtained at the output of the multiplexer (MUX) at a time by applying appropriate logic channel address to the MUX. A software takes care of this and selects the MUX channels automatically. The software developed also takes care of interfacing of the A/D converter with the PC, the data acquisition speed (sampling time of DAS) and the gain of the instrumentation amplifier.

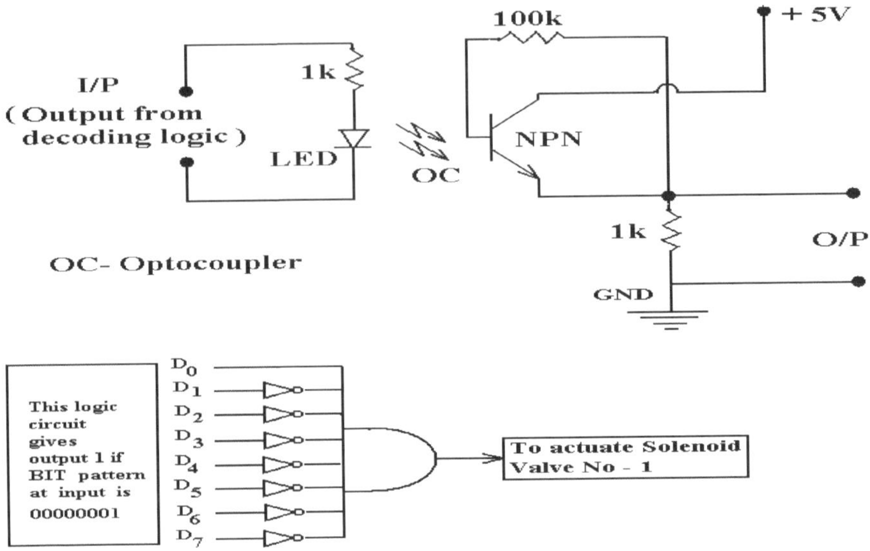

Fig. 2: Isolation and Decoding Logic Circuit

Figure 2 shows the isolation and decoding logic circuit. It isolates and protects the PC control circuit from voltage transients of the actuator circuit through optical isolation.

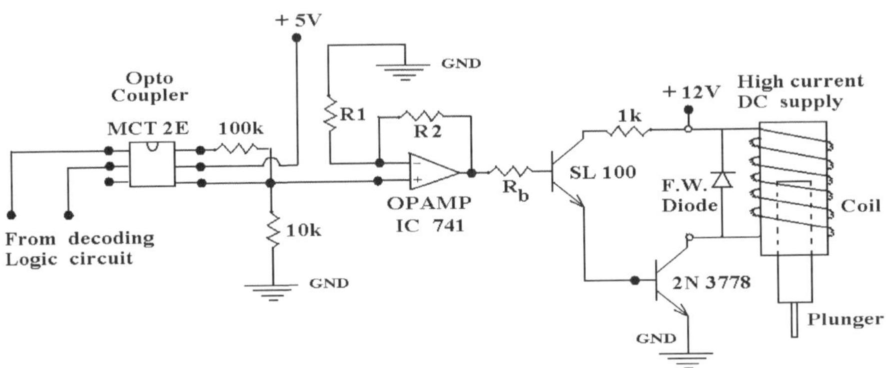

Fig. 3: Hardware Circuit for Plunger Interface

Figure 3 shows hardware circuit for plunger interface with PC. This circuit is driven by a low voltage/current signal which after adequate current amplification drives the high current plunger to actuate different valves.

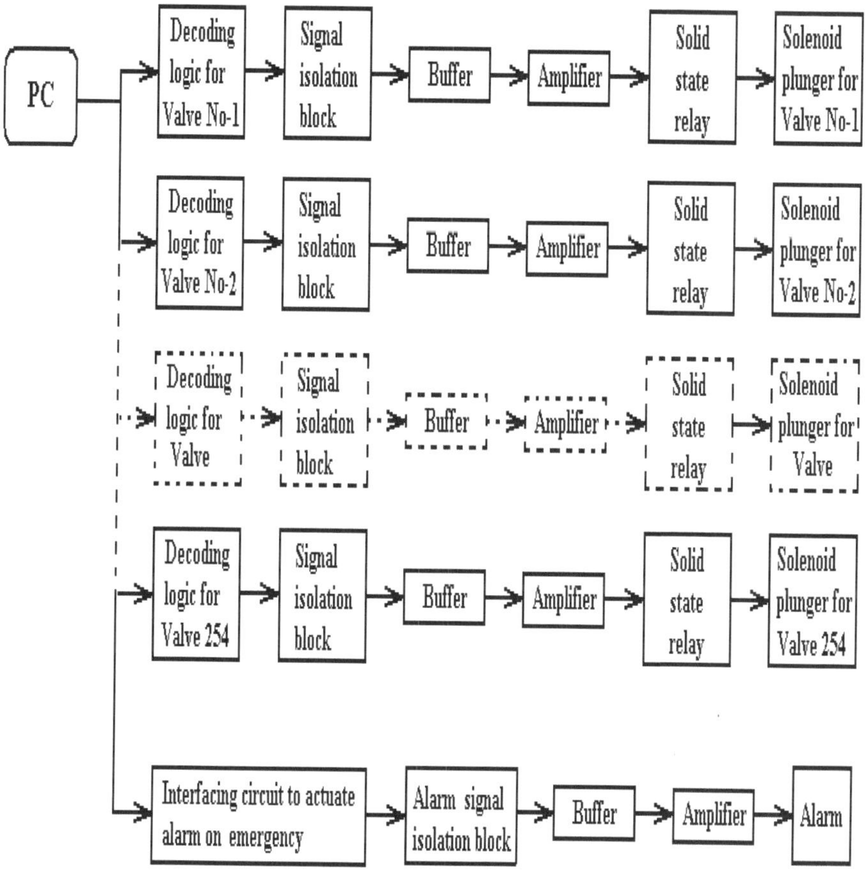

Fig. 4: Solenoid-Plunger Actuation through Parallel Port Using Decoding Logic

Figure 4 shows solenoid-plunger actuation through parallel port of the PC using decoding logic.

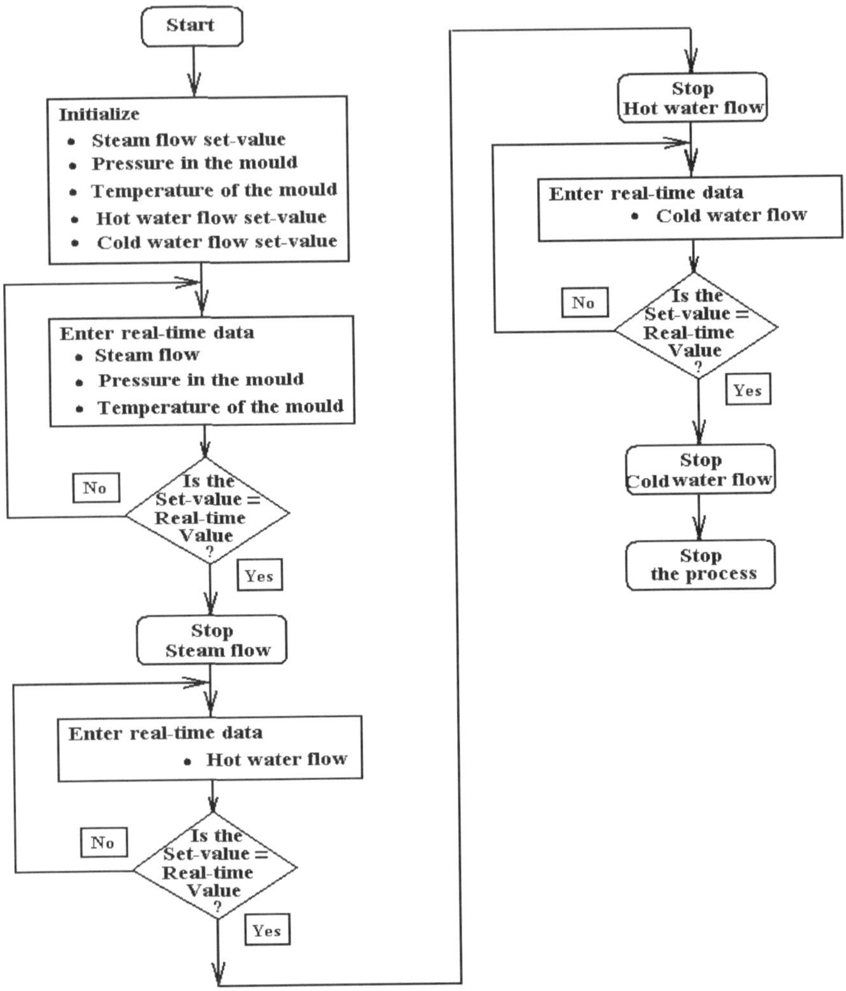

Fig. 5: Algorithm

3. Results

Table 1

Issues	*By Conventional Control*	*By PC-Based Control*
Over Cure Cases	15 out of 100	05 out of 100
Bag-Mark Cases	10 out of 100	03 out of 100
Under Cure Cases	13 out of 100	04 out of 100
Low Pressure Defects	09 out of 100	03 out of 100
Diaphragm Bag Injury	03 out of 100	01 out of 100

4. Conclusions

The proposed on-line PC based control scheme provides extensive cure monitoring. Intricate sensing and subsequent actuation result in significant improvement of over and under cure cases. The on-line control scheme also ensures increased life of diaphragm bags and possible reduction of cure time. The proposed system is also expected to reduce downtime and minimize maintenance problem. Better trace-ability is possible because of PC based control. The powerful software option of the system provides excellent flexibility of control.

5. References

[1] Comb, J.W. (June 1994), "Layered manufacturing control parameters and material selection criteria" Proc. Manufacturing Science and Engineering, ASME, Chicago, USA.

[2] Chattopadhyaya, S. Kumar, Alok Verma " Smart layer approach for built in diagnostic", National Conference on Agile Engineering, Proc. p. 245-251, 1999.

[3] Lin, M., "Composite structures with built-in-diagnostics", Materials today, Volume II, Issue 2, 1999.

[4] "First European Conference on Smart Structures and Materials", Fort Crest Hotel, Glasgow, p.12-14, 1992.

[5] "Sensors and Transducers" D. Patranabis, Prentice-Hall of India (Pvt.) Ltd.

[6] "Transducers and Instrumentation" D.V.S. Murty, Prentice-Hall of India (Pvt.) Ltd.

[7] "Digital Instrumentation" A.J. Bouwens, TATA McGraw-Hill & Co.

[8] "Industrial instrumentation and control"-by S.K. Singh, TATA McGraw-Hill & Co.

Built-in-Diagnosis to Future Technology for Micro-Manufacturing—A Case Study from Tyre Making Plant

Somnath Chattopadhyay[1], Surender Kumar[2] and Pankaj Mishra[2]

[1]Department of Mechanical Engineering and Mining Machinery Engg., Indian School of Mines, Dhanbad-826004

[2]Department of Production Engineering, Birla Institute of Technology, Mesra, Ranchi-835215

1. Introduction

Rapid prototyping system is the newer development in manufacturing technology that involves a group of manufacturing techniques based on layer-by-layer material deposition rather than on material removal or deformation. The technology enables rapid production of complex 3D objects directly from computer aided design model without involving any conventional tooling or numerical control part programming. In intelligent rapid prototyping, the methodologies namely CAD modeling, process planning, fixture design and inspection are intelligently integrated with knowledge-based systems with concurrent flow and control of information between each systems. Moreover, the iterations required to come up with an optimum design could also be performed concurrently and instantly through distributed black board approach in comparison to the time taken in non-concurrent processes of traditional rapid prototyping.

In order to build smart and intelligent systems, the materials must possess some extraordinary properties in comparison to the traditional materials. In a situation related to the loads, mechanical and thermal stresses, the materials themselves have to be proactive to the changing environment. In case of the smart materials, the degrees of intensity of the environmental parameters have to be sensed. If possible, online corrective measures also have to be taken through actuators. In this context, the principle of flexible circuit printing technique have been employed which applies a network of distributed piezo-electric sensors and actuators.

2. Rapid Prototyping

Rapid prototyping is an emerging area of the most promising modern manufacturing technology that involves a group of manufacturing techniques based on the concept of layered material deposition rather than material removal or deformation. Leading companies all over the world are engaged in building model and prototypes for focus group evaluation, testing and down-stream moulding and casting processes through this cutting edge manufacturing methodologies. In true sense rapid prototyping enables a group of users to reduce the cycle time and cost appreciably and causes substantial enhancement of the capability of the designer to optimize the functionality, appeal and quality of the products.

For the last few years, the techniques of rapid prototyping have become commercially available and are now having a significant impact on the overall design-production cycle of the industry. With a strong emphasis on stereolithography, about a thousand of machines came into use by 1995 all over the world and the figure is exponentially increasing around the industrial manufacturing arena where the lion's share of prototyping takes place.

The most recent development in rapid prototyping has been directed towards the development of new materials or techniques of material deposition. Hence, path breaking application based research is needed in different sectors to transform this technology into an efficient, handy and economically affordable approach leading to distinct improvements in the product development life cycle. There are some burning issues as given below and the modern day rapid prototyping technology has to address the same:

- Integration of feature based design and modeling with rapid prototyping technology.
- Enhancement of flexibility and efficiency with intelligent process planning.
- Intelligent fixture design with the improvement in quality.
- Best possible compromise between the conflicting requirements of design and product development.

3. Current Rapid Prototyping Processes

There are six different commercially available rapid prototyping processes that have come out from the research efforts of the universities, research centres and manufacturing companies around the world. These processes are detailed in Table 1 as below.

Table 1: Rapid Prototyping Processes

Process	*Manufacturer*
Stereolithography (SLA)	3D Systems Inc.
Fused Deposition Modeling (FDM)	Stratasys Inc.
Laminated Object Manufacturing (LOM)	Helisys Inc.
Selective Laser Sintering (SLS)	DTM Inc.
Solid Ground Curing (SGC)	Cubital Inc.
3D Plotting	Sanders prototype Inc.

Each of the above technologies involves different methods of generating slices using different types of materials. Several studies have been made to compare and evaluate the different rapid prototyping processes.

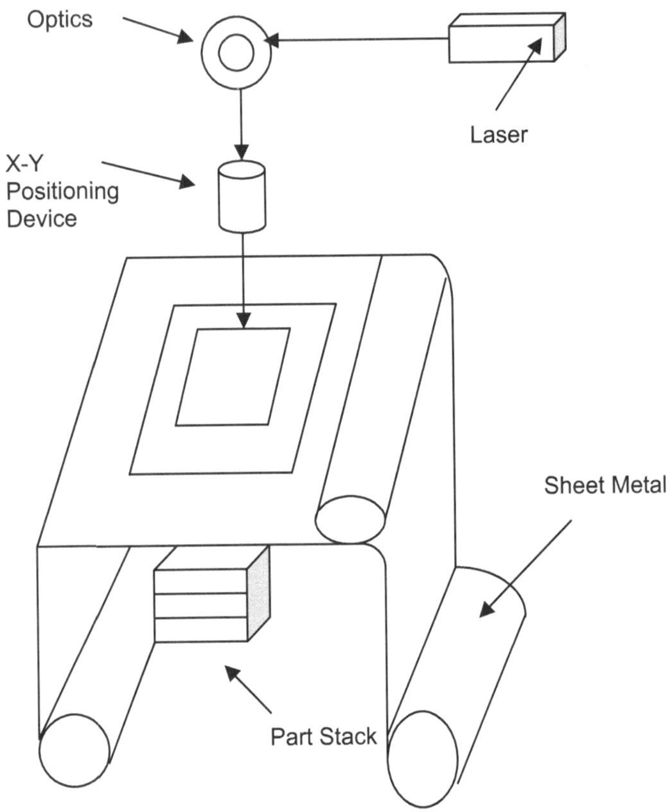

Fig. 1: Laminated Object Manufacturing

Out of the above six processes, the Laminated Object Manufacturing (LOM) technique is suitable for the insertion of sensing capability based smart layer system. The main reason is that the insertion of the piezoelectric based flexible printed circuit that can be easily impregnated into the series of layers manufactured by the laminated object manufacturing process of rapid prototyping. In comparison to LOM, the other rapid prototyping processes undergoing such environmental surround which will either destroy the properties of the smart layer or develop technical difficulty in the insertion in the midst of the manufacturing processes.

4. Laminated Object Manufacturing

Helisys Incorporation developed the process of LOM. It produces prototype by gluing sheets of stack (i.e. paper, plastics and composite). Laminated object manufacturing machine bonds a layer of sheet material (of thickness 0.002″–0.2″) to a stack of previously formed lamination and then a laser beam follows the contour of a cross-section generated by CAD to cut it to the required shape. The sheets are either glued or welded together. The excess materials would support for the next layer.

5. Intelligent Rapid Prototyping

The traditional rapid prototyping process involves a series of isolated and discrete non-interdependent and mutually non-exclusive functions such as computer aided model, process planning to suit part accuracy and formation, designing of fixtures to impart compatibility in part shapes and sizes, inspection and verification, virtual sling of the proposed object of manufacturing, post manufacturing finishing operations. Amongst all such, the computer aided model building, process planning inspection and verification are the most indispensable, crucial and time erosive functions required to manufacture a design-specific and acceptable prototype. All these operations are done sequentially without the aid of computer aided knowledge base and rule guided systems.

The design and successful implementation of such systems would involve the development of several Intelligent Knowledge-Based Systems (IKBS), while each such intelligent system would pursue specific tasks. The development of the integrated environment is oriented to coordinating the activities and information for necessary control and information exchange among these IKBS. This would require adoption and combination of the best possible attributes of the following concepts:

- Feature based modeling concept
- Object-oriented programming concept
- Intelligent knowledge-based concept
- Concurrent engineering concept
- Distributed black board communication and control technologies.

The distributed black board approach can be applied to many complex tasks, when several systems of expertise must interact closely to resolve a complicated problem. A successfully implemented intelligent rapid prototyping system would have the following merits over the traditional systems:

- Higher productivity and greater utilization of resources
- Short-cycled manufacturing
- Substantial reduction in design, planning, fixturing and developmental costs
- Higher accuracy and greater quality of the final product

- Agility in reaction to the ever changing requirements of the highly demanding customer
- Greater access and control of the relevant information at all the levels
- Suitable atmosphere development conducive to cooperative team building.

The major modules of the Intelligent Rapid Prototyping (IRP) system consists of the following Intelligent Knowledge Based Systems (IKBS):

- Intelligent Feature Based Rapid Prototyping Part Designer (IFD)
- Knowledge Based Rapid Prototyping Process Planner (KPP)
- Expert Rapid Prototyping Fixturing Support Designer (EFS)
- A knowledge-Based Value Engineering Inspector (KVE).

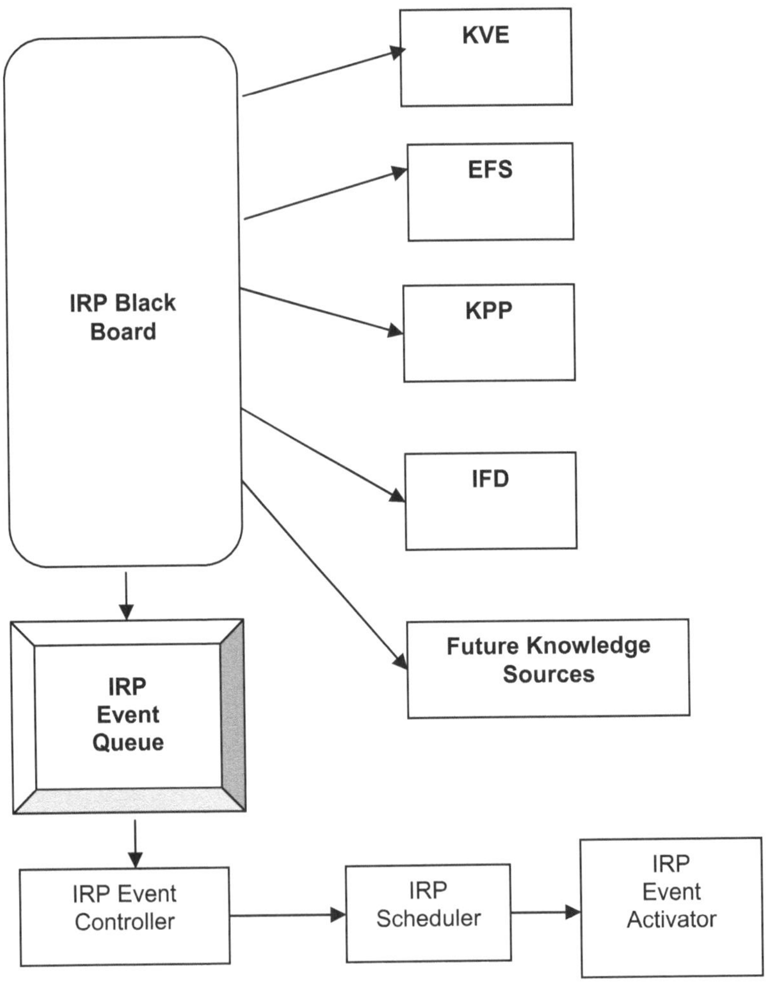

Fig. 2: Configuration of an IRP system

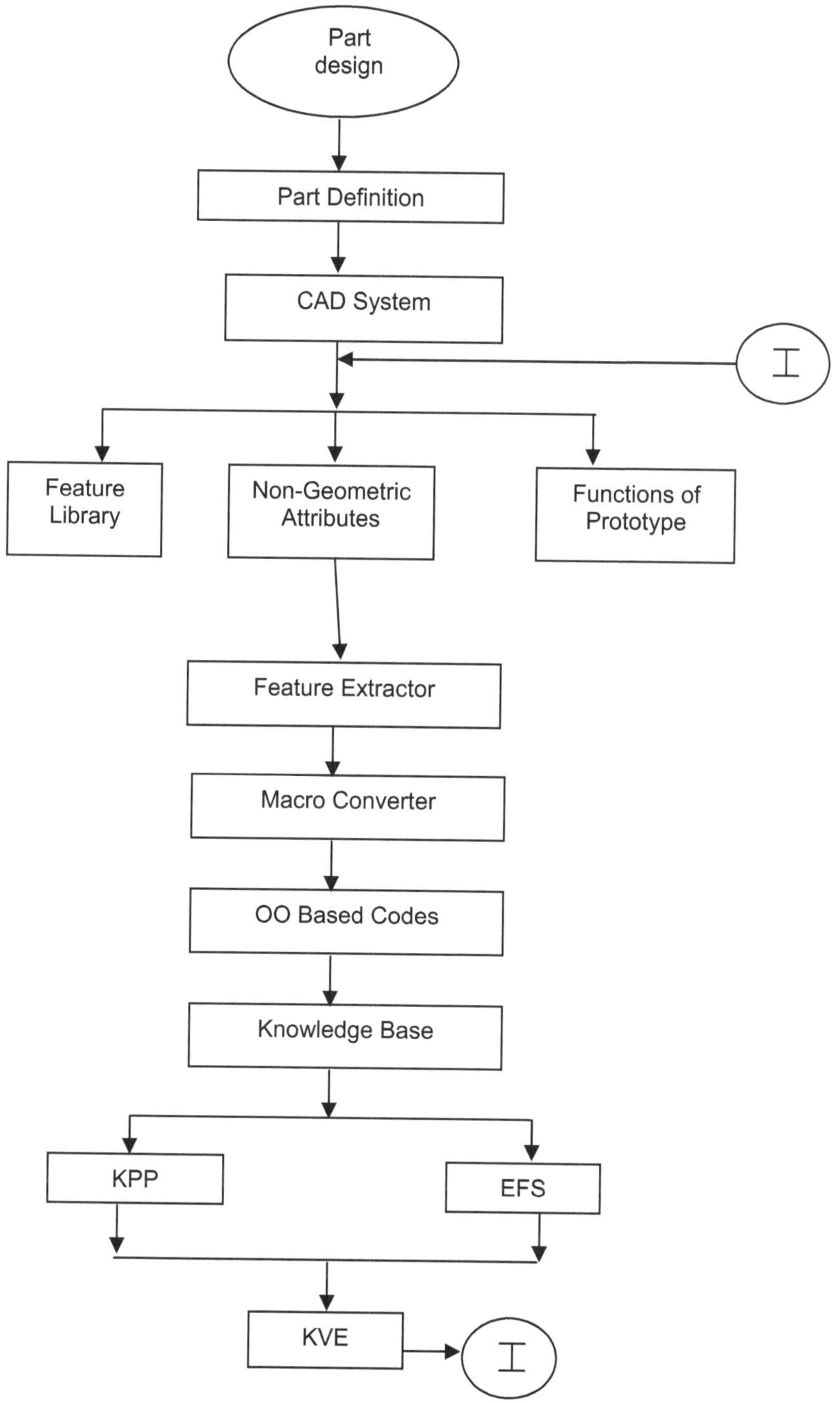

Fig. 3: Operational Outline of IFD Module

6. Smart Layer Approach

There are many sensors that can detect failure in the composite materials. However, implementation of a large network of sensors on new or existing structures would pose immediate challenges for developing a practical monitoring system. Hence, the network of sensors must be easily and reliably integrated with the host structure so that there is no effect on the integrity of structures.

Various methods have been developed following the principle of flexible circuit printing technique to implement a network of distributed piezo-electric sensors/actuators. Such smart flexible layer is made of a dielectric cell with the distributed network of piezo-electric discs serving as sensors and actuators both.

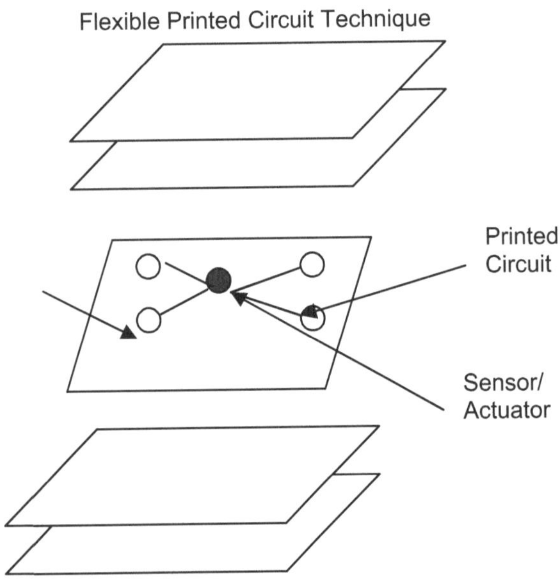

Fig. 4: Schematic view of Smart Layer System

In this context, the main design criteria are given below:

(i) The layer must be able to withstand typical composite cure temperature.

(ii) The layer must provide adequate electric insulation for the embedded wires and devices.

(iii) The layer must have minimal effects on the integrity of the host structures when embedded inside a composite material.

7. Case Study from Tyre Manufacturing Unit

A case study has been done based on a premier Indian rubber manufacturing unit that has two major divisions; one of these is for tyre and the other is for industrial products. Emphasis has been laid on the tyre division in this work as this division provides the bulk of the turnover and profit for the company. After analyzing a set questionnaire for determining the agility of the organization, the following weaknesses of the existing production system have been identified:

Weaknesses

- Comparatively high manufacturing cost
- Inconsistency in product quality
- Comparatively old machineries
- High overhead cost
- Requirement of higher physical labour.

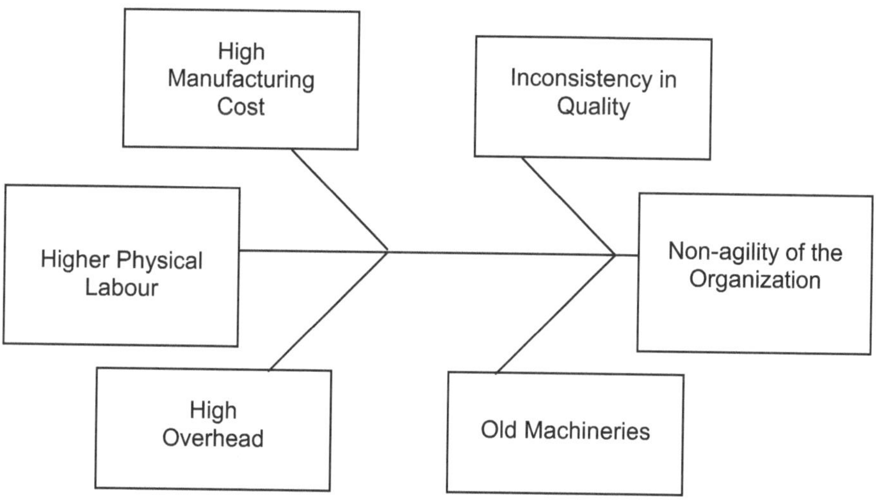

Fig. 5: Fish Bone diagram showing the weaknesses of tyre manufacturing unit

From the above analysis, it is seen that most of the problems in this premier tyre manufacturing company can be reduced substantially through the introduction of intelligent rapid prototyping process in mould making and cure monitoring through the built-in-diagnosis based smart layer in the mould material.

8. Outcome of Technology Implementation

The results of such technology implementation in a rubber tyre manufacturing unit would be mainly reflected in the form of:

- Considerable reduction in cycle time for tyre mould manufacturing.
- Automated process of block mould manufacturing would reduce the dependency on skilled persons resulting considerable reduction in overhead costs.

- Consistency in the accuracy of the mould.
- Pressure cure monitoring can be done effectively.
- Temperature cure monitoring can be implemented to a higher level of accuracy.
- Actuator guided action would result to lesser number of defects in the tyre.
- The defects like bagmark, overcuring zone, undercuring zone, shape deformation are eliminated to a great extent.

The benefits of technology implementation are multifold giving cumulated gainful effects in the subsequent processes of mould making and curing (vulcanizing) for all the downstream operations of the production line namely trimming, finishing, repairing, inspection, testing etc.

Table 2: Post Implementation Results

Sl. No.	*Description*	*Implementation*	
		Before	*After*
1.	Mould making (including design) time	Week	Days
2.	Mould repair time	Days	Hors
3.	Manpower for mould making	4	2
4.	Inspection time after mould making	One shift	1–2 hour(s)
5.	Overcure cases/100 tyres	20	7
6.	Undercure cases/100 tyres	15	5
7.	Bagmark cases/100 tyres	13	3

9. Conclusions

In the light of foregoing discussions, it is revealed that the implementation of rapid prototyping methodology with built-in-diagnostic features will benefit the tyre industry mainly in respect of the following:

- Achieving shortcycled manufacturing for reduced operating cost.
- Intelligent minute level cure monitoring for the micro-manufacturing of rubber products with substantially lesser number of curing defects.

10. References

[1] Rapid prototyping report, CAD/CAM Publishing Inc, 1995.

[2] Hull C. *et al.*, Rapid prototyping current technology and future potential. Rapid Prototyping Journal, 1(1), p. 11–19, 1995.

[3] Aubin, R.F., A worldwide assessment of rapid prototyping technologies. UTRC Report No. 94/13, United Technologies Research Center, East Hartford, USA, 1994.

[4] Rapid Response Manufacturing, Contemporary Methodologies, Tools and Technologies, edited by Jian (John) Dong. Chapman and Hall, 1998.

[5] Comb, J.W. *et al.*, "Layered manufacturing control parameters and material selection criteria" Proc. Manufacturing Science and Engineering, Chicago, ASME, June 1994.

[6] Chattopadhyay, S. *et al.*, "Smart layer approach for built-in-diagnostics", Proc. Of National Conference on Agile Engineering, p. 245-251, 1999.

[7] Lin, M. *et al.*, "Composite structures with built-in-diagnostics" Materials Today, Volume II, Issue 2, 1999.

[8] Proc. Of the First European Conference on Smart Structures and Materials, Glasgow, U.K., p. 12-14, 1992.

Development of OTR Tyre Mould through Reverse Engineering

Vinay Sharma[1], S. Chattopadhyay[2], P.S. Mukherjee[2], Reena Bajaj[2] and Indira Prasanna[2]

[1]Department of Production Engineering, Birla Institute of Technology, Mesra, Ranchi-835215

[2]Department of Mechanical Engineering and Mining Machinery Engineering, Indian School of Mines, Dhanbad-826004

1. Introduction

Manufacturing of the complex OTR moulds is a difficult job. Traditionally, considerable human skill is required to generate the complex profile of the tyre mould. The existing tyre mould making process is time consuming in nature. Due to this reason frequent modifications and changes are not possible in product manufacturing. Reverse engineering is the methodology in which the design is made from the product itself. Stereo-lithography/Three-dimensional scanning/3D Xeroxing can generate complex design from the product directly. Rapid prototyping methodology can produce complex OTR aluminium moulds. The aim of this paper is to highlight the application of reverse engineering technique for making the complex profiles of OTR tyre moulds.

2. Reverse Engineering

Basically, Reverse Engineering is the process of evolution of the technological principles for possible mechanical application through analysis of technological structure and operations. It often involves taking apart a mechanical device or an electronic component or a software program and analyse its function in detail, usually with the intention of constructing a new device or program that performs the same functions without actually copying it directly from the original item.

As Computer Aided Design (CAD) has become more popular, the reverse engineering has become a viable method to create a 3D virtual model of an existing physical component for use in three dimensional computer aided design, Computer Aided Manufacturing (CAM), Computer Aided Engineering (CAE) and others. The reverse engineering process involves the measuring of an object and then reconstructing it as a three dimensional model. The physical object can be measured using three dimensional scanning technologies in Coordinate Measuring Machines, Laser Scaners, White Light Digitizers or Computed Tomography. The measured data, usually represented as a point cloud, lacks topological information and is therefore often processed and modeled into a more usable format such as a triangular faced mesh, a set of Non Uniform Rational B-Splines (NURBS) surface or a CAD model. Applications like Imageware, Rapidform or Geomagic are used to process the point clouds in proper forms to be usable in other applications such as three dimensional CAD, CAM, CAE or visualization.

Reverse engineering is often used in military applications in order to adapt the technology of other nations in respect of devices or parts that have been obtained by the regular troops in the war fields or such items have been acquired by intelligent operations. The technique was often used during the Second World War (WW-II). Well known examples from WW-II include:

- *Jerry Can:* British and American forces noticed that the Germans had gasoline cans with an excellent design. They reverse engineered copies of those cans. The cans were popularly known as *Jery Cans.*
- *Topulev Tu-4:* A number of American B-29 bombers on mission over Japan were forced to land at USSR. The Soviets did not have a similar strategic bomber that time and decided to copy the B-29 bomber. Within a few years a near perfect copy Tu-04 was developed by USSR.
- V2 Rocket technical documents for the V2 and related technologies were captured by the Western Allies at the end of war. Soviet and captured German engineers had to reproduce technical documents and plans, working from captured hardware. In order to make their clone of the rocket i.e. the unit R-1 that began the post-war Soviet rocket program leading to the development of R-7 and the beginning of space race.

3. Three Dimensional Scanning

The development of newer product using the existing product is considered as the start of the approach. Three dimensional scanning facility in reverse engineering has enabled the users to scan the free surface quickly and accurately. High density point clouds that consist of hundreds of thousands of closely packed 3D data points can totally capture the 3D shape to be produced. Each data point consists of x, y and z co-ordinates and the intensity of the surface. With the data point consists of x, y and z coordinates and the intensity of the surface. With the scan alignment software it may be possible to create an align group of scans. The optional modeler package enables:

- Merging of the align scans into one contiguous file for a variety of formats commensurate with industry standards.
- Use of various software packages to manipulate the data.
- To create a 3D CAD file.
- To modify the CAD file according to the requirements of new product.
- To develop a tool path for facilitating the CAM process.
- To produce a new object based on the reversed file.

4. Rapid Prototyping

The Rapid Prototyping (RP) technique would be an important element in the reverse engineering process for possible implementation in a tyre manufacturing unit. It is an emerging area of the most promising modern manufacturing technology. The technique is based on the layered material deposition rather than on material removal or deformation. This form of technology enables rapid manufacturing of complex 3D objects directly from the computer aided design without involving any process planning, traditional tooling or numerically controlled part programming. Leading companies all over the world are engaged in building models and prototypes for evaluation, testing, downstream moulding and casting processes through this cutting edge manufacturing methodology. Rapid prototyping in true sense enables a group of users to reduce the cycle time and cost appreciably and causes substantial enhancement of the capability of the designer to optimize the functionality, appeal and the quality of products. During the last few years, the techniques of rapid prototyping have become commercially available and these techniques are now having a significant impact on the overall design-production cycle of the industry.

The current rapid prototyping based processes are :

- Stereolithography (SLA)
- Fused Deposition Modeling (FDM)
- Laminated Object Manufacturing (LOM)
- Selective Laser Sintering (SLS)
- Solid Ground Curing (SGC)
- 3D Plotting.

With a strong emphasis on stereolithography, thousand of machines came into use by 1995 all over the world and the figure is exponentially increasing around the industrial manufacturing arena where the major share of prototyping takes place. As the most recent development of the rapid prototyping process, it is directed towards the development of new materials or techniques of material deposition. A path breaking application based research is needed in different sectors to transform this technology into an efficient and economically affordable as well as versatile system leading to remarkable improvements in the product life cycle.

5. Reverse Engineering in Tyre Industries

The typical problems faced in the tyre manufacturing units are:

- Comparatively higher physical labour.
- Long cycled manufacturing.
- Comparatively high manufacturing cost.
- Inconsistency in product quality.
- Comparatively old machineries.
- High overhead cost.

The manufacturing of intricate surfaces of the tyre mould is a time taking process and it can be considered as a skill-based operation. Following the conventional methods, it becomes a bottleneck issue to bring out frequent changes in shape and design of the tyre. This leads to non-agility in tyre manufacturing process to go for making abrupt changes in the shape and design of a tyre for satisfying the needs of the customers. The typical tyre moulds consist of the different conjugate grooves and profiles for the generation of ribs and lugs of the tyre. It becomes a cumbersome task to generate such kind of intricate grooves in the mould. These operations depend considerably on the human skill and thus it becomes very much time consuming. Hence, an extensive feature-based library can be prepared out of the existing rib and lug patterns and then through rapid prototyping different complicated aluminium moulds can be manufactured easily in no time. For this purpose, aluminium based Laminated Object Manufacturing system would be ideally suitable.

Basically, in reverse engineering the design is generated from the product itself. The complex profile of the object is generated from the tested product and not from the idea or any brain storming session. Thus, it is an intelligent copying system of the tested design. However, smart and flexible profile generation is the essential requirement in reverse engineering process. Further, the information technology can play an important role in implementing the reverse engineering process.

The implementation of Reverse Engineering in tyre manufacturing would be beneficial in the following ways:

- Complex design profiles of the tyre moulds can be manufactured in least possible time with reduced operating cost.
- The designer would be able to go for wider options in manufacturing as compared to that submitted by the other competitors.
- Less labour intensive and less dependent on individual human skill.

Table 1 presents the summarized benefits in the application of reverse engineering for tyre manufacturing.

Table 1

Sl. No.	*Problem*	*Before*	*After*
1.	New mould design through reverse engineering	Weeks	Hours
2.	Design-to-manufacture cycle	Month	Week
3.	Manpower for mould making	4	2
4.	Inspection time for tyre moulds	One shift	Minutes
5.	Internet based manufacturing	Not possible	Possible

Index

B

Bajaj, Reena 151, 178
Banerjee, S. 128
Basu, A.K. 151

C

Chattopadhyay, S. 151, 161, 168, 178
Choudhary, M.K. 115
Chowdhury, Tanushree 124

D

Darad, R.K. 25
Das, U. 83

E

Emranuzzaman 115

G

George, Jacob 133
Ghosh, R. 25
Gupte, S.S. 78, 94

J

Jagdish 115
Jain, V.K. 65

K

Kumar, Rajesh 1
Kumar, Surender 168

M

Mahapatra, W.P. 83
Mishra, N.M. 124
Mishra, Pankaj 168
Mittra, D.K. 161
Mukherjee, K. 25
Mukherjee, P.S. 178
Muthreja, I.L. 78, 94

N

Nabiullah 115

P

Pal, Kaushik 100
Pal, Samir K. 100
Pandey, A.K. 51
Pingua, B.M.P. 115
Prasanna, Indira 151, 178

R

Rai, R.L. 51
Rao, P.T. 11

S

Sadhu, P.K. 161
Sengupta, R. 141
Sengupta, R.P. 1
Sharma, Vinay 178
Shukla, Sandeep 51
Singh, J.N. 51
Singh, S.N. 51
Singh, Sureshwar 69
Singh, V.B. 146

T

Thangavelu, N. 38
Tiwary, P.N. 69
Tripathi, R.N. 51

V

Vidyarthi, Deepak 58

W

Waldia, M.S. 128